CAMBRIDGE CHECKPOINTS 2009

VCE Further Mathematics

- Questions from past examinations
- Core facts, skills and extended response tasks
- Fully worked solutions
- Online quizzes

Neil Duncan

CAMBRIDGE UNIVERSITY PRESS
Cambridge, New York, Melbourne, Madrid, Cape Town, Singapore, São Paulo

Cambridge University Press
477 Williamstown Road, Port Melbourne, VIC 3207, Australia

www.cambridge.edu.au
Information on this title: www.cambridge.org/9780521739481
© Neil Duncan 2008
First published 2008

Cover design by Julia Bojanovic
Printed in Australia by BPA Print Group Ltd

For cataloguing data please visit the Libraries Australia website:
http://librariesaustralia.nla.gov.au

Cambridge Checkpoints VCE Further Mathematics
ISBN 978-0-521-73948-1

Reproduction and Communication for educational purposes
The Australian *Copyright Act 1968* (the Act) allows a maximum of one chapter or 10% of the pages of this publication, whichever is the greater, to be reproduced and/or communicated by any educational institution for its educational purposes provided that the educational institution (or the body that administers it) has given a remuneration notice to
Copyright Agency Limited (CAL) under the Act.

For details of the CAL licence for educational institutions contact:
Copyright Agency Limited
Level 15, 233 Castlereagh Street
Sydney NSW 2000
Telephone: (02) 9394 7600
Facsimile: (02) 9394 7601
Email: info@copyright.com.au

Reproduction and Communication for other purposes
Except as permitted under the Act (for example a fair dealing for the purposes of study, research, criticism or review) no part of this publication may be reproduced, stored in a retrieval system, communicated or transmitted in any form or by any means without prior written permission. All inquiries should be made to the publisher at the address above.

Cambridge University Press has no responsibility for the persistence or accuracy
of URLs for external or third-party internet websites referred to in this
publication and does not guarantee that any content on such websites is, or will remain, accurate or appropriate.

All Victorian Curriculum and Assessment Authority material copyright VCAA. Reproduced by kind permission of the Victorian Curriculum and Assessment Authority Victoria, Australia.

Disclaimer:
This book has been independently published by Cambridge University Press for use by teachers and students. Although material contained has been reproduced with the permission of the VCAA, this publication is not connected with nor endorsed by them in any way.

VCE examination papers for the current year are reproduced from the papers used by students during their examinations. However, at the time of this book's publication, the VCAA had not yet officially released the VCE examination papers to the public. Final corrections, amendments and/or alterations to the papers could still be made and the final official papers will then be posted on the VCAA website, which can be found at:
http://www.vcaa.vic.edu.au/vce/exams/index.html

Every effort has been made to trace and acknowledge copyright. The publishers apologise for any accidental infringement and welcome information that would rectify any error or omission in subsequent editions.

Contents

Introduction iv
Further Mathematics Formula Sheet v

Core facts and skills tasks

A1	Core: univariate data	1
A2	Core: correlation and regression	11
A3	Core: time series..	17

Modules

B1	Number patterns and applications..	25
B2	Geometry and trigonometry	35
B3	Graphs and relations	43
B4	Business-related mathematics	55
B5	Networks and decision mathematics	67
B6	Matrices..	77

Extended responses

C1	Core – extended responses..	85
C2	Number patterns – extended responses.................................	96
C3	Geometry and trigonometry – extended responses	104
C4	Graphs and relations – extended responses...........................	116
C5	Business-related mathematics – extended responses	128
C6	Networks and decision mathematics – extended responses.......	139
C7	Matrices – extended responses..	151

Solutions

A1	Core: univariate data	156
A2	Core: correlation and regression	160
A3	Core: time series..	163
B1	Number patterns and applications..	167
B2	Geometry and trigonometry	173
B3	Graphs and relations	179
B4	Business-related mathematics	186
B5	Networks and decision mathematics	193
B6	Matrices..	197
C1	Core – extended responses..	201
C2	Number patterns – extended responses.................................	206
C3	Geometry and trigonometry – extended responses	213
C4	Graphs and relations – extended responses...........................	223
C5	Business-related mathematics – extended responses	231
C6	Networks and decision mathematics – extended responses.......	239
C7	Matrices – extended responses..	243

Introduction

Revision and preparation for the assessment tasks, conducted under test conditions, is essential. A selection of material taken from past test papers for Further Mathematics is presented for you here.

The multiple-choice material is in the first two sections of this book. It has been broken up into a compulsory or 'core' section (part A), and modules (part B). Which modules you take will depend on your particular circumstances.

There are also longer questions from the modules. Please note that the answers given throughout this book are guideline answers only. They were written by the author and in no way represent 'official answers'. Every attempt was made to ensure their accuracy.

Some modifications to the Study Design for Further Mathematics were made in 2006. For instance, the module Matrices was new for 2006 and had no prior examination paper examples to present here. The matrices 'sample questions' have been removed for 2009, returning this book to '100% past examination questions' status.

Answer the analysis tasks with care and neatness. Use words as well as numbers. Explain what you are doing. On the assumption that you may make an error somewhere, seek the best marks for what you have done. Go for uncluttered clarity! Spread your work out, number your questions (and parts) clearly and use your best writing.

This edition of **Cambridge Checkpoints VCE Further Mathematics** includes Study Cards as an appendix. These are designed to be used during the year, and especially in the days and hours leading up to the exams. They are in the form of a compete list of the topics within the course, with some key formulae for each.

It is suggested that these cards be cut from the text and used regularly throughout the year to ensure these critical points are investigated fully. They are an essential point of reference for your final exam preparation.

Best of luck for the year!

Further Mathematics Formula Sheet

Written examinations 1 and 2

Core: Data analysis

standardised score: $z = \dfrac{x - \bar{x}}{s_x}$

least squares line: $y = a + bx$ where $b = r\dfrac{s_y}{s_x}$ and $a = \bar{y} + b\bar{x}$

residual value: residual value = actual value − predicted value

seasonal index: seasonal index = $\dfrac{\text{actual figure}}{\text{deseasonalised figure}}$

Module 1: Number patterns and applications

arithmetic series: $a + (a+d) + \ldots + (a+(n-1)d) = \dfrac{n}{2}[2a + (n-1)d] = \dfrac{n}{2}(a+l)$

geometric series: $a + ar + ar^2 + \ldots + ar^{n-1} = \dfrac{a(1-r^n)}{1-r}, r \neq 1$

infinite geometric series: $a + ar + ar^2 + ar^3 \ldots = \dfrac{a}{1-r}, |r| < 1$

Module 2: Geometry and trigonometry

area of a triangle: $\dfrac{1}{2}bc\sin A$

Heron's formula: $A = \sqrt{s(s-a)(s-b)(s-c)}$ where $s = \dfrac{1}{2}(a+b+c)$

circumference of a circle: $2\pi r$
area of a circle: πr^2
volume of a sphere: $\dfrac{4}{3}\pi r^3$
surface area of a sphere: $4\pi r^2$
volume of a cone: $\dfrac{1}{3}\pi r^2 h$
volume of a cylinder: $\pi r^2 h$
volume of a prism: area of base × height
volume of a pyramid: $\dfrac{1}{3}$ area of base × height
Pythagoras' theorem: $c^2 = a^2 + b^2$
sine rule: $\dfrac{a}{\sin A} = \dfrac{b}{\sin B} = \dfrac{c}{\sin C}$
cosine rule: $c^2 = a^2 + b^2 - 2ab\cos C$

Module 3: Graphs and relations

Straight line graphs

gradient (slope): $m = \dfrac{y_2 - y_1}{x_2 - x_1}$

equation: $y = mx + c$

Module 4: Business-related mathematics

simple interest: $I = \dfrac{PrT}{100}$

compound interest: $A = PR^n$, where $R = 1 + \dfrac{r}{100}$

hire purchase: effective rate of interest $\approx \dfrac{2n}{n+1} \times$ flat rate

Module 5: Network and decision mathematics

Euler's formula: $v + f = e + 2$

Module 6: Matrices

determinant of a 2×2 matrix: $A = \begin{bmatrix} a & b \\ c & d \end{bmatrix}$; $\det A = \begin{vmatrix} a & b \\ c & d \end{vmatrix} = ad - bc$

inverse of a 2×2 matrix: $A^{-1} = \dfrac{1}{\det A} \begin{bmatrix} d & -b \\ -c & a \end{bmatrix}$ where $\det A \neq 0$

A1. Core: univariate data

The following information relates to questions 1 and 2.
The percentage investment returns of seven superannuation funds for the year 2002 are
–4.6%, –4.7%, 2.9%, 0.3%, –5.5%, –4.4%, –1.1%.

Question 1

The median investment return is

A. –4.7%   **B.** –4.6%   **C.** –4.5%   **D.** –4.4%   **E.** 0.3%

[VCAA 2003 FM]

Question 2

The range of investment returns is

A. 2.6%   **B.** 3.5%   **C.** 4.0%   **D.** 5.5%   **E.** 8.4%

[VCAA 2003 FM]

Question 3

The distribution of test scores obtained when 2500 students sit for an examination is bell-shaped with a mean of 64 and a standard deviation of 12. From this information we can conclude that the number of these students who obtained marks between 52 and 76 is closest to

A. 68   **B.** 95   **C.** 850   **D.** 1700   **E.** 2375

[VCAA 2003 FM]

The following information relates to questions 4 and 5.
The mean weight of twelve people is 72 kg; the standard deviation of the weights of these twelve people is 5 kg.

Question 4

The total weight of the twelve people is

A. 77 kg   **B.** 360 kg   **C.** 864 kg   **D.** 924 kg   **E.** 4320 kg

[VCAA 2003 FM]

A1. Core: univariate data

Question 5

These twelve people are about to go on a rafting adventure. Before boarding the raft, they are all required to put on a life-saving vest that weighs 2 kg. The effective weight of each person is now their weight plus the weight of the life-saving vest. The effective weights of the twelve people have

A. a mean of 72 kg with a standard deviation of 5 kg.
B. a mean of 72 kg with a standard deviation of 7 kg.
C. a mean of 74 kg with a standard deviation of 5 kg.
D. a mean of 74 kg with a standard deviation of 7 kg.
E. a mean of 74 kg with a standard deviation of 10 kg.

[VCAA 2003 FM]

The following information relates to questions 6 and 7.

The level of water usage of 250 houses was rated in a survey as low, medium or high and the size of the houses as small, standard or large. The results of the survey are displayed in the table below.

Level of water usage	Size of house		
	small	standard	large
low	15	14	9
medium	22	71	11
high	15	47	46

Question 6

The percentage of standard sized houses rated as having a high level of water usage is

A. 18.8%   B. 35.6%   C. 43.5%   D. 47.0%   E. 53.8%

[VCAA 2003 FM]

Question 7

The variables, **level of water usage** and **size of house**, as recorded in this survey, are

A. both numerical variables.
B. both categorical variables.
C. neither numerical nor categorical variables.
D. numerical and categorical variables respectively.
E. categorical and numerical variables respectively.

[VCAA 2003 FM]

A1. Core: univariate data

The following information relates to questions 8 and 9.

The marks obtained by students who sat for a test are displayed as an ordered stemplot as shown.

```
0 | 9
1 |
2 | 0 1 2 5 6
3 | 0 1 1 1 3 5 5 7 8 9 9
4 | 1 2 3 4 4 6 7 7
5 | 0
```

Question 8

The number of students who sat the test is

A. 25   **B.** 26   **C.** 27   **D.** 32   **E.** 50

[VCAA 2004 FM]

Question 9

The interquartile range of these test marks is closest to

A. 9   **B.** 13   **C.** 30   **D.** 36   **E.** 41

[VCAA 2004 FM]

Question 10

The distribution of the weights of eggs produced by a chicken farm is approximately bell-shaped with a mean of 85 g and a standard deviation of 5 g. Eggs weighing 95 g or more are classified as Extra Large. The percentage of eggs that would be classified as Extra Large is closest to

A. 0.15%   **B.** 0.35%   **C.** 2.5%   **D.** 5%   **E.** 16%

[VCAA 2004 FM]

The following information relates to questions 11 and 12.

The number of DVD players in each of 20 households is recorded in the frequency table below.

Number of DVD players	Frequency
0	6
1	9
2	3
3	1
4	0
5	1
	20

A1. Core: univariate data

Question 11

For this sample of households, the percentage of households with **at least** one DVD player is

A. 30% B. 45% C. 50% D. 70% E. 90%

[VCAA 2004 FM]

Question 12

For this sample of households, the mean number of DVD players in these 20 households is

A. 0.75 B. 1.00 C. 1.15 D. 1.64 E. 2.00

[VCAA 2004 FM]

The following information relates to questions 13 and 14.

The level of Internet usage (never used, sometimes used, often used) for 217 school students sampled from Years 3 to 12 is indicated in the table on the next page. Some of the entries in the table are missing.

Level of Internet usage	3-6	Year group 7-10	11-12	Total
never used	44	9	8	
sometimes used	16			58
often used	10		47	
Total			73	217

Question 13

For this sample of students, the total number of students who never used the Internet is

A. 44 B. 51 C. 61 D. 70 E. 217

[VCAA 2004 FM]

Question 14

The percentage of Year 7–10 students who sometimes used the Internet is closest to

A. 11% B. 24% C. 27% D. 28% E. 32%

[VCAA 2004 FM]

A1. Core: univariate data

The following information relates to questions 15 and 16.

The table below lists the speed (in km/h) of ten cars recorded in a 60 km/h zone. Also recorded are the ages (in years) of the drivers.

Speed	Age
71.8	27
68.3	38
65.1	22
63.2	64
62.8	57
62.6	37
62.5	21
61.3	19
60.1	57
59.8	61

Question 15

The median speed (in km/h) of the ten cars is

A. 62.6   **B.** 62.7   **C.** 62.8   **D.** 63.0   **E.** 63.5

[VCAA 2005 FM]

Question 16

The percentage of the drivers over the age of 25 years is

A. 30%   **B.** 40%   **C.** 50%   **D.** 60%   **E.** 70%

[VCAA 2005 FM]

Question 17

The histogram shown is best described as

A. negatively skewed.
B. positively skewed.
C. symmetric.
D. negatively skewed with outliers.
E. positively skewed with outliers.

[VCAA 2005 FM]

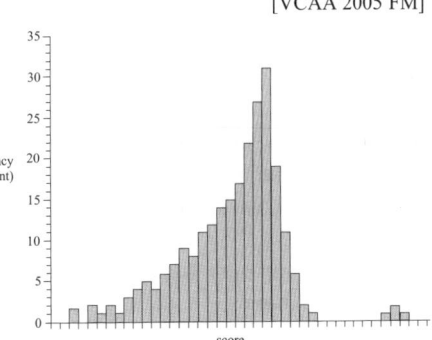

5

A1. Core: univariate data

The following information relates to questions 18 and 19.

Text messaging use (never, sometimes, everyday) and the number of mobile phones in the household were recorded for a sample of 154 households. The results are shown in the table below.

Text messaging use	Number of mobile phones in household				
	0	1	2	3	Total
Never	34	10	3	0	47
Sometimes	0	23	12	2	37
Everyday	0	45	15	10	70
Total	34	78	30	12	154

Question 18

Of the households with two mobile phones in the sample, the percentage that never used text messaging is

A. 0% B. 6% C. 10% D. 20% E. 30%

[VCAA 2005 FM]

Question 19

The mean number of mobile phones in these 154 households is closest to

A. 1.13 B. 1.45 C. 1.50 D. 1.54 E. 2.00

[VCAA 2005 FM]

Question 20

The distribution of fuel consumption of a particular model of car is approximately bell-shaped with a mean of 8.8 km per litre and a standard deviation of 2.2 km per litre. The percentage of this model of car that has a fuel consumption less than 6.6 km per litre is closest to

A. 2.5% B. 5% C. 16% D. 32% E. 68%

[VCAA 2005 FM]

Question 21

As part of an experiment, three samples of pine trees were planted. Each sample contained 50 trees. One sample was grown under hot conditions, one sample was grown under mild conditions and one sample was grown under cool conditions.

The parallel box plots (next page) show the rate of growth (in centimetres per year) of these three samples.

A1. Core: univariate data

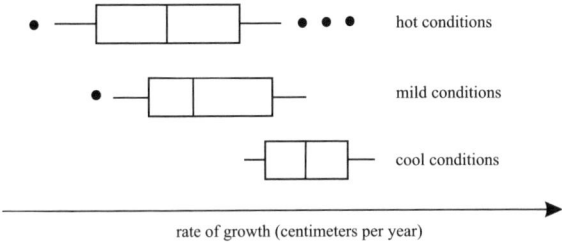

From the parallel box plots it can be concluded that, as conditions change from hot to mild to cool, the rate of growth for these trees

A. decreases on average and becomes less variable.
B. decreases on average and becomes more variable.
C. does not change on average but becomes more variable.
D. increases on average and becomes less variable.
E. increases on average and becomes more variable.

[VCAA 2005 FM]

The following information relates to questions 22, 23 and 24.

The back-to-back ordered stemplot below shows the distribution of maximum temperatures (in °Celsius) of two towns, Beachside and Flattown, over 21 days in January.

```
        Beachside        Flattown
            9 8 7 5  | 1 |  8 9
      4 3 2 2 1 1 0 0| 2 |
            9 9 8 7 6 5| 2 |  8 9
                   3 2| 3 |  3 3 4
                     8| 3 |  5 5 6 7 7 7 8 8
                      | 4 |  0 0 1 2
                      | 4 |  5 6
```

Question 22

The variables temperature (°Celsius) and town (Beachside or Flattown) are

A. both categorical variables.
B. both numerical variables.
C. categorical and numerical variables respectively.
D. numerical and categorical variables respectively.
E. neither categorical nor numerical variables.

[VCAA 2006 FM]

A1. Core: univariate data

Question 23

For **Beachside**, the range of maximum temperatures is

A. 3°C   B. 23°C   C. 32°C   D. 33°C   E. 38°C

[VCAA 2006 FM]

Question 24

The distribution of maximum temperatures for **Flattown** is best described as

A. negatively skewed.
B. positively skewed.
C. positively skewed with outliers.
D. approximately symmetric.
E. approximately symmetric with outliers.

[VCAA 2006 FM]

Question 25

The head circumference (in cm) of a population of infant boys is normally distributed with a mean of 49.5 cm and a standard deviation of 1.5 cm. Four hundred of these boys are selected at random and each boy's head circumference is measured. The number of these boys with a head circumference of less than 48.0 cm is closest to

A. 3   B. 10   C. 64   D. 272   E. 336

[VCAA 2006 FM]

The following information relates to questions 26 and 27.

The distribution of test marks obtained by a large group of students is displayed in the percentage frequency histogram to the right.

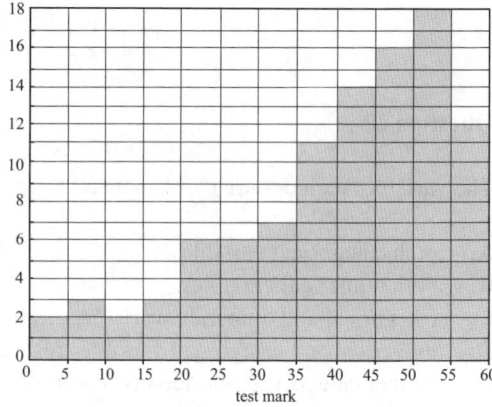

Question 26

The pass mark on the test was 30 marks. The percentage of students who passed the test is

A. 7%   B. 22%
C. 50%   D. 78%
E. 87%

[VCAA 2006 FM]

A1. Core: univariate data

Question 27

The median mark lies between

A. 35 and 40   **B.** 40 and 45   **C.** 45 and 50   **D.** 50 and 55   **E.** 55 and 60

[VCAA 2006 FM]

The following information relates to questions 28 and 29.

The dot plot below shows the distribution of the number of bedrooms in each of 21 apartments advertised for sale in a new high-rise apartment block.

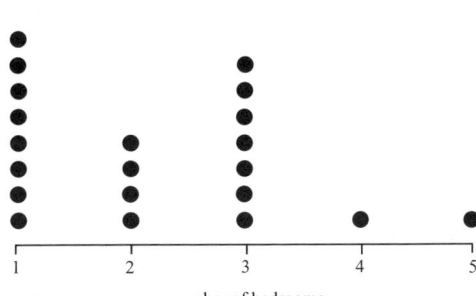

Question 28

The mode of this distribution is

A. 1   **B.** 2   **C.** 3   **D.** 7   **E.** 8

[VCAA 2007 FM]

Question 29

The median of this distribution is

A. 1   **B.** 2   **C.** 3   **D.** 7   **E.** 8

[VCAA 2007 FM]

Question 30

A student obtains a mark of 56 on a test for which the mean mark is 67 and the standard deviation is 10.2. The student's standardised mark (standard z-score) is closest to

A. −1.08   **B.** −1.01   **C.** 1.01   **D.** 1.08   **E.** 49.4

[VCAA 2007 FM]

Question 31

The length of 3-month-old baby boys is approximately normally distributed with a mean standard deviation of 1.6 cm. The percentage of 3-month-old baby boys with length greater than 59.5 cm is closest to

A. 5%   **B.** 16%   **C.** 68%   **D.** 84%   **E.** 95%

[VCAA 2007 FM]

9

A1. Core: univariate data

The following information relates to questions 32 and 33.

Samples of jellyfish were selected from two different locations, A and B. The diameter (in mm) of each jellyfish was recorded and the resulting data is summarised in the boxplots shown here.

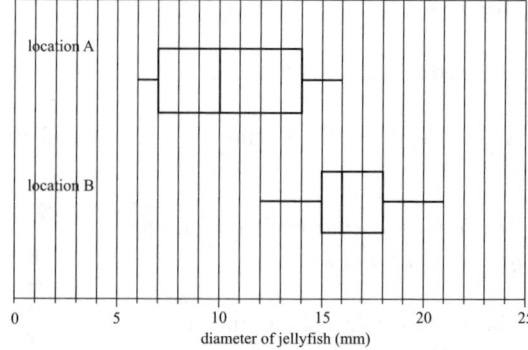

Question 32

The percentage of jellyfish taken from location A with a diameter greater than 14 mm is closest to

A. 2% B. 5% C. 25% D. 50% E. 75%

[VCAA 2007 FM]

Question 33

From the boxplots, it can be concluded that the diameters of the jellyfish taken from location A are generally

A. similar to the diameters of the jellyfish taken from location B.
B. less than the diameters of the jellyfish taken from location B and less variable.
C. less than the diameters of the jellyfish taken from location B and more variable.
D. greater than the diameters of the jellyfish taken from location B and less variable.
E. greater than the diameters of the jellyfish taken from location B and more variable.

[VCAA 2007 FM]

Question 34

The relationship between the variables
    size of car (1 = small, 2 = medium, 3 = large)
and
    salary level (1 = low, 2 = medium, 3 = high)
is best displayed using

A. a scatterplot.
B. a histogram.
C. parallel boxplots.
D. a back-to-back stemplot.
E. a percentaged segmented bar chart.

[VCAA 2007 FM]

A2. Core: correlation and regression

The following information relates to questions 35 and 36.

Eighteen students sat for a 15 question multiple-choice test. In the scatterplot shown here, the number of errors made by each student on the test is plotted against the time they reported studying for the test. A least squares regression line has been determined for this data and is also displayed on the scatterplot.

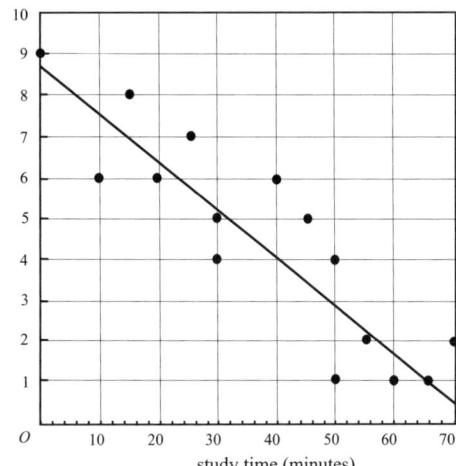

The equation for the least squares regression line is

number of errors = 8.8 − 0.120 × study time

and the coefficient of determination is 0.8198.

Question 35

Using the least squares regression line, it can be estimated that, on average, a student reporting a study time of 35 minutes would make

A. 4.3 errors.   B. 4.6 errors.   C. 4.8 errors.
D. 5.0 errors.   E. 13.0 errors.

[VCAA 2003 FM]

Question 36

The value of Pearson's product-moment correlation coefficient, r, for this data, correct to two decimal places, is

A. −0.91   B. −0.82   C. 0.67   D. 0.82   E. 0.91

[VCAA 2003 FM]

A2. Core: correlation and regression

Question 37

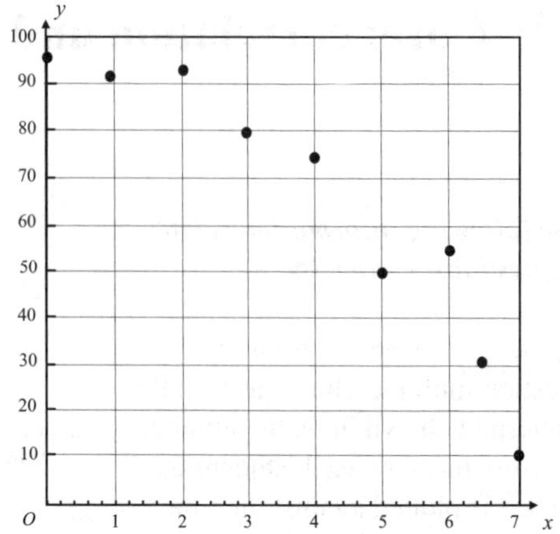

The relationship between the two variables y and x, as shown in the scatterplot, is nonlinear.

Which one of the following transformations, by itself, is most likely to linearise this data?

A. a $\frac{1}{x}$ transformation.

B. a $\frac{1}{y}$ transformation y.

C. an x^2 transformation.   **D.** a log x transformation.   **E.** a log y transformation.

[VCAA 2003 FM]

Question 38

The relationship between **resting pulse rate** (in beats per minute) and **fitness level** (below average, average, above average) is best displayed using

A. a histogram   **B.** a scatterplot   **C.** a time series plot.
D. parallel boxplots   **E.** back-to-back stemplots.

[VCAA 2003 FM]

The following information relates to questions 39 and 40.

The average rainfall and temperature range at several different locations in the South Pacific region are displayed in the scatterplot.

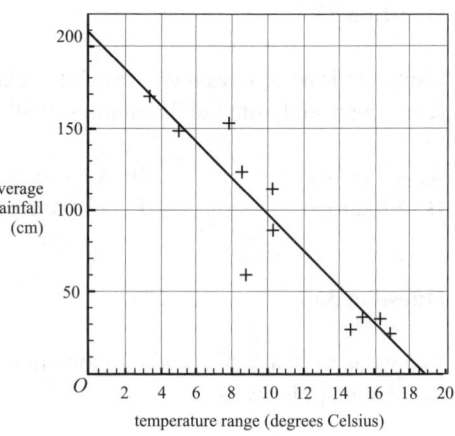

Question 39

A least squares regression line has been fitted to the data as shown. The equation of this line is closest to

A. average rainfall = 210 − 11 × temperature range.
B. average rainfall = 210 + 11 × temperature range.
C. average rainfall = 18 − 0.08 × temperature range.
D. average rainfall = 18 + 0.08 × temperature range.
E. average rainfall = 250 − 13 × temperature range.

[VCAA 2004 FM]

12

A2. Core: correlation and regression

Question 40

The value of the product moment correlation coefficient, r, for the data, is $r = -0.9260$. The value of the coefficient of determination is

A. -0.9260 B. -0.8575 C. 0.8575 D. 0.9260 E. 0.9623

[VCAA 2004 FM]

The following information relates to questions 41 and 42.

The time spent batting and the number of runs scored in a cricket player's last nine games are plotted on the scatterplot shown here.

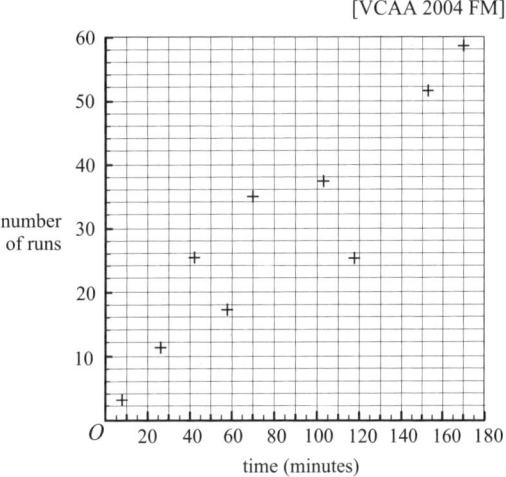

Question 41

When a three median regression line is fitted to the scatterplot, its slope is closest to

A. 0.2 B. 0.3
C. 0.4 D. 0.5
E. 0.6

[VCAA 2004 FM]

Question 42

The data point (70, 55) should have been plotted instead of the point (70, 35). If this mistake is corrected, the slope of a redrawn three median regression line will be

A. very much greater than before. B. greater than before.
C. less than before. D. very much less than before.
E. unchanged.

[VCAA 2004 FM]

The following information relates to questions 43 and 44.

The length (in metres) and wingspan (in metres) of eight commercial airplanes are displayed in the table below.

Length	70.7	70.7	63.7	58.4	54.9	39.4	36.4	33.4
Wingspan	64.4	59.6	60.3	60.3	47.6	35.8	28.9	28.9

A2. Core: correlation and regression

Question 43

Correct to four decimal places, the value of Pearson's product moment correlation coefficient for this data is

A. 0.9371   B. 0.9583   C. 0.9681   D. 0.9793   E. 0.9839

[VCAA 2005 FM]

Question 44

The equation of the least squares regression line for this data is
$$\text{wingspan} = -2.99 + 0.96 \times \text{length}$$
From this equation it can be concluded that, on average, for these airplanes, wingspan

A. decreases by 2.03 metres with each one metre increase in length.
B. increases by 0.96 metres with each one metre increase in length.
C. decreases by 0.96 metres with each one metre increase in length.
D. increases by 2.99 metres with each one metre increase in length.
E. decreases by 2.99 metres with each one metre increase in length.

[VCAA 2005 FM]

Question 45

For a set of bivariate data, involving the variables x and y,
$r = -0.5675$, $\bar{x} = 4.56$, $s_x = 2.61$, $\bar{y} = 23.93$, and $s_y = 6.98$.
The equation of the least squares regression line $y = a + bx$ is closest to

A. $y = 30.9 - 1.52x$ B. $y = 17.0 - 1.52x$ C. $y = -17.0 + 1.52x$
D. $y = 30.9 - 0.2x$ E. $y = 24.9 - 0.2x$

[VCAA 2006 FM]

Question 46

The waist measurement (cm) and weight (kg) of 12 men are displayed in the table below.

waist (cm)	84	74	89	75	106	114	80	101	101	94	126	82
weight (kg)	84	72	67	59	97	112	67	91	98	89	117	62

Using this data, the equation of the least squares regression line that enables weight to be predicted from waist measurement is
$$\text{weight} = -20 + 1.11 \times \text{waist}$$
When this equation is used to predict the weight of the man with a waist measurement of 80 cm, the residual value is closest to

A. −11 cm   B. 11 cm   C. −2 cm   D. 2 cm   E. 69 cm

[VCAA 2006 FM]

A2. Core: correlation and regression

Question 47

A student uses the following data to construct the scatterplot shown below.

x	0	1	2	3	4	5	6	7	8	9
y	5	7	14	33	58	76	124	166	188	238

To linearise the scatterplot, she applies an x-squared transformation. She then fits a least squares regression line to the transformed data with y as the dependent variable. The equation of this least squares regression line is closest to

A. $y = 7.1 + 2.9x^2$
B. $y = -29.5 + 26.8x^2$
C. $y = 26.8 - 29.5x^2$
D. $y = 1.3 + 0.04x^2$
E. $y = -2.2 + 0.3x^2$

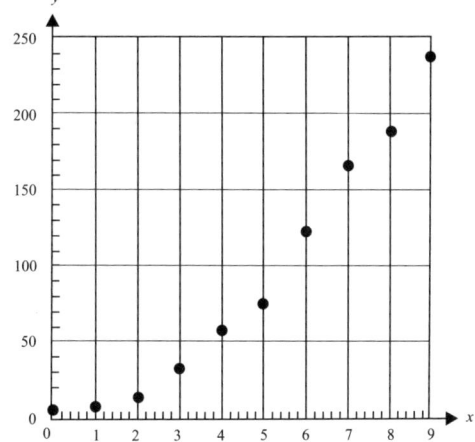

[VCAA 2006 FM]

The following information relates to Questions 48 and 49.

The lengths and diameters (in mm) of a sample of jellyfish selected from another location were recorded and displayed in the scatterplot shown here. The least squares regression line for this data is shown. The equation of the least squares regression line is
length = 3.5 + 0.87 × diameter

The correlation coefficient is $r = 0.9034$

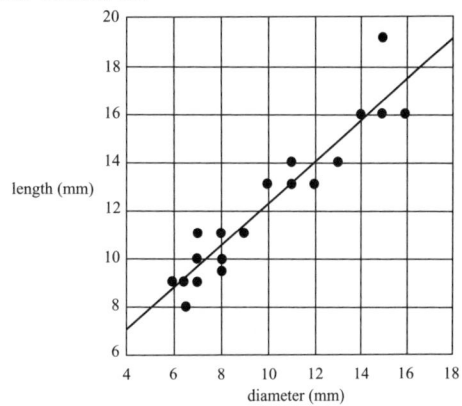

Question 48

Written as a percentage. the coefficient of determination is closest to

A. 0.816%   B. 0.903%   C. 81.6%   D. 90.3%   E. 95.0%

[VCAA 2007 FM]

15

A2. Core: correlation and regression

Question 49

From the equation of the least squares regression line, it can be concluded that for these jellyfish, on average

A. there is a 3.5 mm increase in *diameter* for each 1 mm increase in *length*.
B. there is a 3.5 mm increase in *length* for each 1 mm increase in *diameter*.
C. there is a 0.87 mm increase in *diameter* for each 1 mm increase in *length*.
D. there is a 0.87 mm increase in *length* for each 1 mm increase in *diameter*.
E. there is a 4.37 mm increase in *diameter* for each 1 mm increase in *length*.

[VCAA 2007 FM]

Question 50

A student uses the following data to construct the scatterplot shown below.

x	1	2	3	4	5	6	7	8	9
y	12	25	33	58	98	168	345	397	869

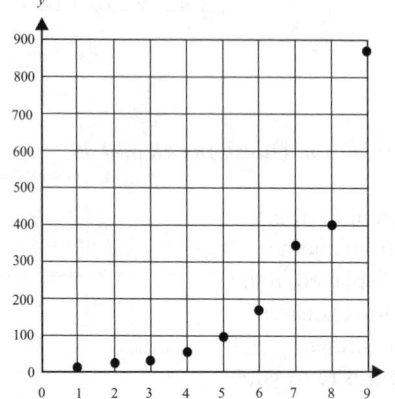

To linearise the scatterplot, she applies a **log y** transformation; that is, a log transformation is applied to the y-axis scale.

She then fits a least squares regression line to the transformed data. With x as the independent variable, the equation of this least squares regression line is closest to

A. $\log y = -217 + 88.0x$
B. $\log y = -3.8 + 4.4x$
C. $\log y = 3.1 + 0.008x$
D. $\log y = 0.88 + 0.23x$
E. $\log y = 1.58 + 0.002x$

[VCAA 2007 FM]

A3. Core: time series

The following relates to questions 51 and 52.

A time series plot of the maximum daily temperature over a nine-day period is shown.

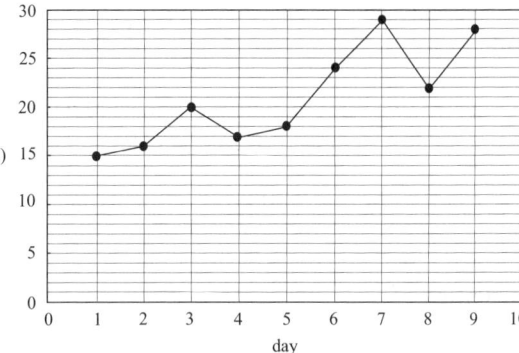

Question 51

If the three-median technique is used to fit a trend line to this time series, its slope (in °C per day) is closest to

A. 1 **B.** 2 **C.** 3 **D.** 4 **E.** 5

[VCAA 2000 FM]

Question 52

After fitting a three-median trend line to this time series, it was found that the maximum temperature on day 3 was 30°C, not 20°C. If this correction is made to the time series plot and the three-median trend line refitted to the data, its slope will be

A. much greater than before. **B.** a little greater than before.
C. the same as before. **D.** a little less than before.
E. much less than before.

[VCAA 2000 FM]

Question 53

Umbrella sales for 1999–2000 and their seasonal indices are shown in the table below.

	Summer	Autumn	Winter	Spring
Number of umbrellas sold	6 786	7 891	11 897	9 564
Seasonal index	0.72	0.81	1.33	1.14

The deseasonalised value of the number of umbrellas sold in winter (to the nearest whole number) is

A. 8945 **B.** 9034 **C.** 9425 **D.** 11 897 **E.** 15 823

[VCAA 2000 FM]

A3. Core: time series

Question 54

The monthly sales of red wine (in thousands of litres) is shown for 1982.

Month	Jan.	Feb.	Mar.	Apr.	May	June	July	Aug.	Sept.	Oct.	Nov.	Dec.
Sales (thousands of litres)	464	675	703	887	1139	1077	1318	1260	1120	963	996	960

Using the two-term moving mean method **with** centring, the smoothed value for wine sales in May, in thousands of litres, is

A. 951.5   B. 1013   C. 1060.5   D. 1108   E. 1139

[VCAA 2001 FM]

Question 55

Red wine sales for the period 1982–1987 are shown in the time series plot here.
From this plot it can be said that red wine sales for the period 1982–1987 showed

A. seasonality only.
B. an increasing trend with seasonality.
C. a decreasing trend only.
D. three-yearly cycles.
E. random variation only.

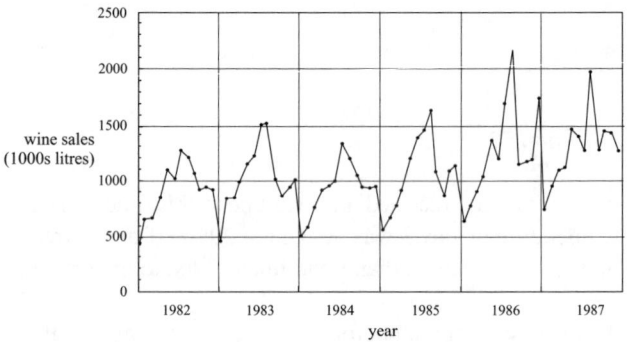

[VCAA 2001 FM]

Question 56

The time series shown here charts the change in interest rates over a period of years.

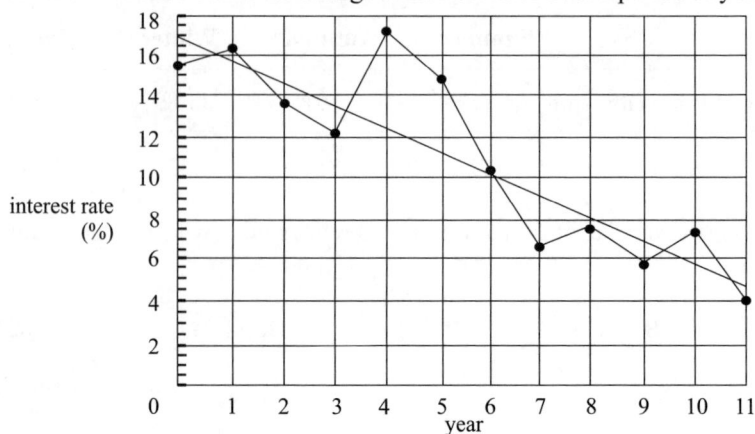

A3. Core: time series

A trend line has been fitted to the data as shown. The equation of the trend line is closest to

A. rate $= -1.1 \times$ year.
B. rate $= 15.6 - 1.5 \times$ year.
C. rate $= 17 + 1.1 \times$ year.
D. rate $= -1.1 + 17 \times$ year.
E. rate $= 17 - 1.1 \times$ year.

[VCAA 2002 FM]

Question 57

The data below gives the number of accidents recorded at a city intersection each year from 1993 to 2002.

Year	Number of accidents
1993	13
1994	7
1995	3
1996	9
1997	10
1998	8
1999	7
2000	6
2001	10
2002	11

Using a **four point** moving average (mean) with centring, the smoothed value of the number of accidents in 1995 is

A. 7.25
B. 7.375
C. 7.5
D. 7.625
E. 8

[VCAA 2003 FM]

Question 58

The seasonal indices for the first three quarters of a year are shown in the table below.

	Quarter 1	Quarter 2	Quarter 3	Quarter 4
Seasonal index	1.05	0.84	0.92	

The seasonal index for Quarter 4 is

A. 0.88
B. 0.94
C. 1.00
D. 1.08
E. 1.19

[VCAA 2003 FM]

A3. Core: time series

Question 59

The time series plot shows the share price of two companies over a period of time. From the plot, it can be concluded that over the interval 1990–2000, the **difference** in share price between the two companies has shown

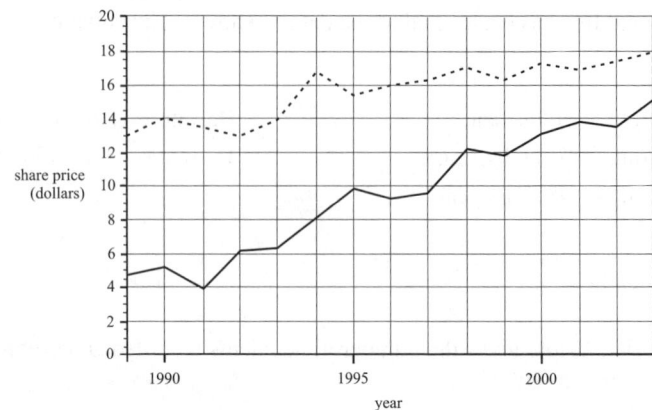

A. a decreasing trend.
D. a five-year cycle.
B. an increasing trend.
E. no trend.
C. seasonal variation.

[VCAA 2004 FM]

Question 60

The quarterly seasonal indices for mineral water sales (in litres) of a mineral water supplier are shown in the table below.

	Quarter 1	Quarter 2	Quarter 3	Quarter 4
Seasonal index	1.28	1.02	0.74	0.96

When deseasonalised, the amount of mineral water sold in Quarter 1 is 28 098 litres. To the nearest litre, the actual amount of mineral water sold in Quarter 1 was

A. 7025 litres.
D. 35 965 litres.
B. 21 952 litres.
E. 112 392 litres.
C. 28 098 litres.

[VCAA 2004 FM]

Question 61

The five points shown on the grid have been taken from a time series plot that is to be smoothed using median smoothing.

The coordinates of the median of these five points are

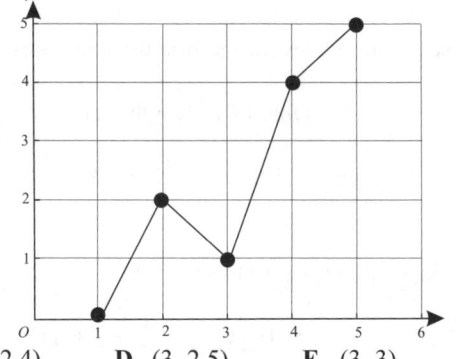

A. (3, 1)   B. (3, 2)   C. (3, 2.4)   D. (3, 2.5)   E. (3, 3)

[VCAA 2005 FM]

A3. Core: time series

Question 62

The quarterly seasonal indices for sales in a shop are shown in the table below.

Quarter	1	2	3	4
Seasonal index	1.3	0.9	0.7	1.1

A seasonal index of 1.1 for Quarter 4 means that sales in Quarter 4 are typically

A. 10% above the yearly average.
B. 10% below the yearly average.
C. 11% below the yearly average.
D. 90% above the yearly average.
E. 90% below the yearly average.

[VCAA 2005 FM]

The following information relates to questions 63 and 64.

The month-by-month price of a share listed on the Australian Stock Exchange is shown in the time series plot for a 36-month period. Also shown is a least squares regression line that has been fitted to the data. The equation of the least squares regression line is: share price = 1.24 + 0.06 × month

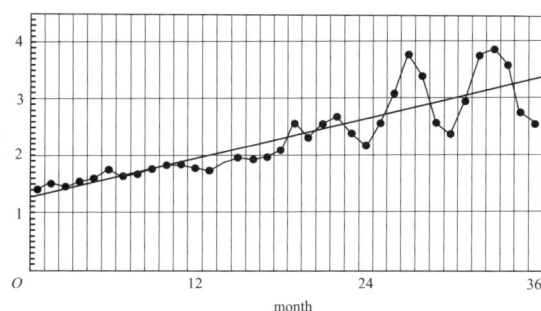

Question 63

The least squares regression line predicts that the price of the share after 48 months will be

A. $1.96   B. $4.12   C. $5.04   D. $28.80   E. $62.40

[VCAA 2005 FM]

Question 64

Which one of the following statements **best** describes the time series plot for the period shown?

A. The share price shows no trend and no change in variability.
B. The share price shows no trend and increases in variability.
C. The share price shows an increasing linear trend with constant variability.
D. The share price shows an increasing linear trend with decreasing variability.
E. The share price shows an increasing linear trend with increasing variability.

[VCAA 2005 FM]

A3. Core: time series

Question 65

The table below displays the total monthly rainfall (in mm) in a reservoir catchment area over a one-year period.

month	Jan	Feb	Mar	Apr	May	June	July	Aug	Sept	Oct	Nov	Dec
rainfall	9	65	35	99	75	90	133	196	106	56	76	76

Using three mean moving average smoothing, the smoothed value for the total rainfall in April is closest to

A. 65   B. 66   C. 70   D. 75   E. 88

[VCAA 2006 FM]

The following information relates to questions 66, 67 and 68.

month	Jan	Feb	Mar	Apr	May	June	July	Aug	Sept	Oct	Nov	Dec
seasonl index	1.30	1.21	1.00	0.95	0.95	0.86	0.86	0.89	0.94		0.99	1.07

The table shows the seasonal indices for the monthly unemployment numbers for workers in a regional town.

Question 66

The seasonal index for October is missing from the table. The value of the missing seasonal index for October is

A. 0.93   B. 0.95   C. 0.96   D. 0.98   E. 1.03

[VCAA 2006 FM]

Question 67

The actual number of unemployed in the regional town in September is 330. The **deseasonalised** number of unemployed in September is closest to

A. 310   B. 344   C. 351   D. 371   E. 640

[VCAA 2006 FM]

A3. Core: time series

Question 68

A trend line that can be used to forecast the deseasonalised number of unemployed workers in the regional town for the first nine months of the year is given by

deseasonalised number of unemployed = 373.3 − 3.38 × month number

where month 1 is January, month 2 is February, and so on. The **actual** number of unemployed for June is predicted to be closest to

A. 304   B. 353   C. 376   D. 393   E. 410

[VCAA 2006 FM]

The following information relates to questions 69, 70 and 71.

The time series plot below shows the revenue from sales in dollars) each month made by a Queensland souvenir shop over a three-year period.

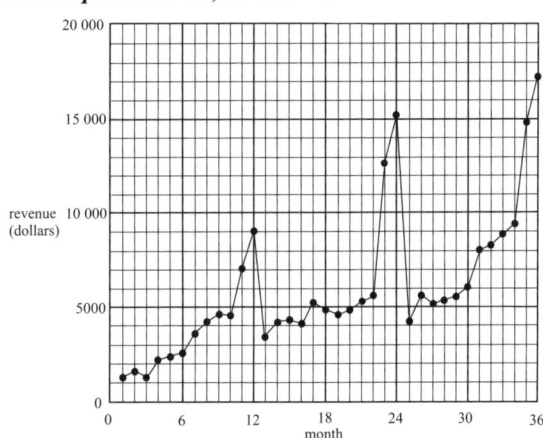

Question 69

This time series plot indicates that, over the three-year period, revenue from sales each month showed

A. no overall trend.
B. no correlation.
C. positive skew.
D. an increasing trend only.
E. an increasing trend with seasonal variation.

[VCAA 2007 FM]

Question 70

A three median trend line is fitted to this data. Its slope (in dollars per month) is closest to

A. 125   B. 146   C. 167   D. 188   E. 255

[VCAA 2007 FM

A3. Core: time series

Question 71

The revenue from sales (in dollars) each month for the first year of the three-year period is shown below.

Month	Revenue ($)
January	1236
February	1567
March	1240
April	2178
May	2308
June	2512
July	3510
August	4234
September	4597
October	4478
November	7034
December	8978

If this information is used to determine the seasonal index for each month, the seasonal index for **September** will be closest to

A. 0.80   B. 0.82   C. 1.16   D. 1.22   E. 1.26

[VCAA 2007 FM]

B1. Number patterns and applications

Question 72

For the sequence 4, 10, 16, 22, ... the sum of the first ten terms is

A. 52   B. 58   C. 310   D. 340   E. 620

[VCAA 2003 FM]

Question 73

In an arithmetic sequence, the second term is 36 and the fourth term is 20. The first term is

A. 20   B. 28   C. 44   D. 52   E. 56

[VCAA 2003 FM]

Question 74

A large pile of bricks is stored at a building site. To make the pile more stable, the bottom layer has 47 bricks, the second layer has 43 bricks, the third layer has 39 bricks and so on. If this pattern continues, the number of bricks in the 11th layer is

A. 3   B. 5   C. 7   D. 9   E. 11

[VCAA 2003 FM]

Question 75

Which one of the following sequences is **not** a geometric sequence?

A. 1, 0.1, 0.01, 0.001, ...   B. 1, 1.1, 1.01, 1.001, ...   C. 3, 3, 3, 3, ...

D. 16, -8, 4, -2, ...   E. $4, 4^2, 4^3, 4^4$, ...

[VCAA 2003 FM]

Question 76

A tank contains 18 000 litres of water. The wall cracks and water flows out at a rate of 120 litres per minute. Three hours later, the amount of water left in this tank is

A. 0 litres.   B. 3 600 litres.   C. 7 200 litres.
D. 10 800 litres.   E. 14 400 litres.

[VCAA 2003 FM]

B1. Number patterns and applications

Question 77

A shrub, 20 cm high, was planted in a pot. After it was planted in the pot, its height increased by 8 cm in the first month, by 4 cm in the second month and by 2 cm in the third month. Assuming that this pattern of growth continues, the shrub will grow to a maximum height of

A. 35 cm   B. 36 cm   C. 37 cm   D. 38 cm   E. 40 cm

[VCAA 2003 FM]

Question 78

A sequence is described by the difference equation $t_{n+1} = 0.4t_n$ where $t_1 = 100$. Which one of the following best describes the sequence?

A. a sequence which is not a geometric sequence
B. a decreasing geometric sequence with all positive terms
C. an increasing geometric sequence with all positive terms
D. a geometric sequence with alternating positive and negative terms
E. a decreasing geometric sequence with negative terms later in the sequence

[VCAA 2003 FM]

Question 79

The first five terms of a sequence of numbers are 20, 10, 20, 10, 20,... A difference equation that generates this sequence is

A. $t_{n+1} = 20 - t_n$ B. $t_{n+1} = t_n - 20$ C. $t_{n+1} = 0.5t_n$
D. $t_{n+1} = t_n - 10$ E. $t_{n+1} = 30 - t_n$

[VCAA 2003 FM]

Question 80

The first three terms of a geometric sequence are 10, 7 and 4.9. The fourth term in this sequence is

A. 3.43   B. 3.47   C. 3.7   D. 4.0   E. 4.2

[VCAA 2004 FM]

Question 81

The sum of the infinite geometric sequence −5.4, 1.8, −0.6, 0.2, ... is

A. −4.04   B. −4.05   C. −4.15   D. −8.0   E. −8.1

[VCAA 2004 FM]

B1. Number patterns and applications

Question 82

The number of goats on a farm is increasing by 8% per annum. At the start of 2003 there were 600 goats on the farm. At the start of 2005, the number of goats on the farm will be closest to

A. 616 B. 648 C. 696 D. 700 E. 756

[VCAA 2004 FM]

The following information relates to questions 83 and 84.

In the first week of training Sophia swims five laps of the pool each day. In the second week she swims seven laps each day, in the third week she swims nine laps each day, and so on.

Question 83

In the 12th week of training the number of laps that Sophia swims each day is

A. 21 B. 23 C. 25 D. 27 E. 31

[VCAA 2004 FM]

Question 84

Sophia swims seven days each week. Assuming this pattern of training continues, the total number of laps of the pool she has swum after 20 weeks is

A. 301 B. 420 C. 960 D. 2940 E. 3360

[VCAA 2004 FM]

Question 85

A sequence follows the rule $w_{n+1} = 4w_n + 2$ where w_n is the nth term and $n = 1, 2, 3, 4...$ The value of the second term, w_2, is 10. The value of the fourth term, w_4, is

A. 18 B. 20 C. 42 D. 170 E. 200

[VCAA 2004 FM]

Question 86

A sequence is defined by the difference equation $t_{n+1} = t_n - 5$ where $t_1 = 15$. The nth term of the sequence is given by

A. $t_n = n - 20$
B. $t_n = 10 - 5n$
C. $t_n = 15 - n$
D. $t_n = 15 - 5n$
E. $t_n = 20 - 5n$

[VCAA 2004 FM]

27

B1. Number patterns and applications

Question 87

Consider the arithmetic sequence: −5, 1, 7. Another number that also appears in this sequence is

A. 15 B. 17 C. 19 D. 21 E. 23

[VCAA 2005 FM]

Question 88

Consider the following collection of sequences.
1.1, 1.11, 1.111, 1.1111 ...
3, 6, 12, 24...
$\frac{1}{2}, \frac{1}{4}, \frac{1}{8}, \frac{1}{16}$ ...
27, 9, 3, 1 ...

The number of **geometric** sequences is

A. 0 B. 1 C. 2 D. 3 E. 4

[VCAA 2005 FM]

Question 89

Gino paid $1200 to hire a hall last year. Gino hired the same hall this year. The ratio of the amount paid to hire the hall last year to the amount paid to hire the hall this year was 3:5. The amount Gino paid to hire the hall this year was

A. $450 B. $720 C. $750 D. $2000 E. $3200

[VCAA 2005 FM]

Question 90

A child has a box of 50 blocks. She lays them out in block towers, starting with tower 1, following the pattern as shown here.
If she continues to follow the same pattern, the maximum number of block towers that she can make is

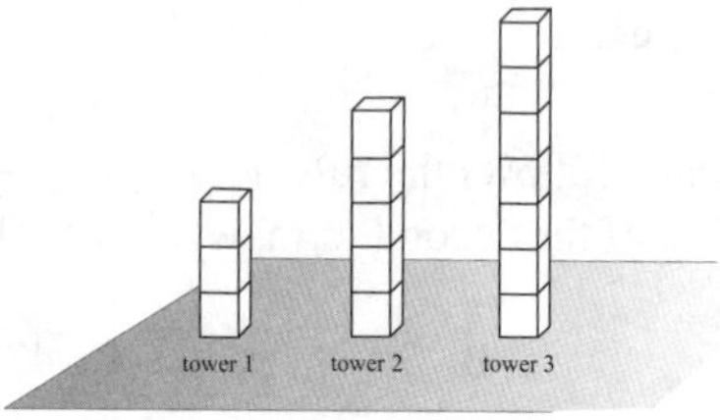

A. 4 B. 5 C. 6 D. 7 E. 24

[VCAA 2005 FM]

B1. Number patterns and applications

Question 91

The first term of an infinite geometric sequence is 80. Each successive term is 75% of the value of the previous term. The sum of the infinite sequence is

A. 80 B. 106 C. 140 D. 240 E. 320

[VCAA 2005 FM]

Question 92

The first term in a sequence is 3. The second term is obtained by multiplying the first term by 5, then subtracting 2. The third term is obtained by multiplying the value of the second term by 5, then subtracting 2. This pattern continues. The difference equation which generates this sequence is

A. $t_n = 5t_{n+1} - 2$ where $t_1 = 3$
B. $t_{n+1} = 5t_n - 2$ where $t_1 = 3$
C. $-5t_{n+1} = t_n - 2$ where $t_1 = 3$
D. $5t_{n+1} = t_n - 2$ where $t_1 = 3$
E. $t_{n+1} = -5t_n - 2$ where $t_1 = 3$

[VCAA 2006 FM]

Question 93

The graph shows the first six terms of a sequence. This sequence is **best** described as

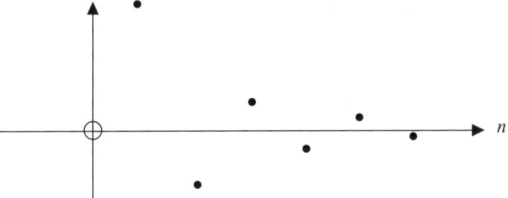

A. geometric with $-1 < r < 0$
B. geometric with $0 < r < 1$
C. geometric with $r > 1$
D. arithmetic with $d < 0$
E. arithmetic with $0 < d < 1$

[VCAA 2005 FM]

Question 94

A sequence is generated from the difference equation $t_{n+1} = at_n + b$ where $t_1 = 3$. The sequence is arithmetic if

A. $a = -1, b = 0$ B. $a = -1, b = 1$ C. $a = 1, b = 2$
D. $a = 2, b = 0$ E. $a = 2, b = -1$

[VCAA 2005 FM]

B1. Number patterns and applications

Question 95

The rule for a difference equation is $P_n = 2P_{n-1} - 200$. If $P_6 = 1000$, then P_4 is equal to

A. 100 B. 400 C. 600 D. 1800 E. 3400

[VCAA 2005]

Question 96

Which one of the following sequences shows the first five terms of an arithmetic sequence?

A. 1, 3, 9, 27, 81...
B. 1, 3, 7, 15, 31...
C. −10, −5, 5, 10, 15...
D. −4, −1, 2, 5, 8...
E. 1, 3, 8, 15, 24...

[VCAA 2006 FM]

Question 97

The first three terms of a geometric sequence are 6, x, 54. A possible value of x is

A. 9 B. 15 C. 18 D. 24 E. 30

[VCAA 2006 FM]

The following information relates to Questions 98 and 99.

A farmer plans to breed sheep to sell. In the first year she starts with 50 breeding sheep. During the first year, the sheep numbers increase by 84%. At the end of the first year, the farmer sells 40 sheep.

Question 98

How many sheep does she have at the start of the second year?

A. 2 B. 42 C. 52 D. 84 E. 92

[VCAA 2006 FM]

Question 99

If S_n is the number of sheep at the start of year n, a difference equation that can be used to model the growth in sheep numbers over time is

A. $S_{n+1} = 1.84 S_n - 40$ where $S_1 = 50$
B. $S_{n+1} = 0.84 S_n - 50$ where $S_1 = 40$
C. $S_{n+1} = 0.84 S_n - 40$ where $S_1 = 50$
D. $S_{n+1} = 0.16 S_n - 50$ where $S_1 = 40$
E. $S_{n+1} = 0.16 S_n - 40$ where $S_1 = 50$

[VCAA 2006 FM]

B1. Number patterns and applications

Question 100

A difference equation is defined by $f_{n+1} - f_n = 5$ where $f_1 = -1$.
The sequence $f_1, f_2, f_3, \ldots$ is

A. 5, 4, 3, ...
B. 4, 9, 14, ...
C. $-1, -6, -11, \ldots$
D. $-1, 4, 9, \ldots$
E. $-1, 6, 11, \ldots$

[VCAA 2006 FM]

Question 101

A crystal measured 12.0 cm in length at the beginning of a chemistry experiment. Each day it increased in length by 3%. The length of the crystal after 14 days' growth is closest to

A. 12.4 cm B. 16.7 cm C. 17.0 cm D. 17.6 cm E. 18.2 cm

[VCAA 2006 FM]

Question 102

The values of the first five terms of a sequence are plotted on the graph shown here.

The first order difference equation that could describe the sequence is

A. $t_{n+1} = t_n + 5 \quad t_1 = 4$
B. $t_{n+1} = 2t_n + 1 \quad t_1 = 4$
C. $t_{n+1} = t_n - 3 \quad t_1 = 4$
D. $t_{n+1} = t_n + 3 \quad t_1 = 4$
E. $t_{n+1} = 3t_n \quad t_1 = 4$

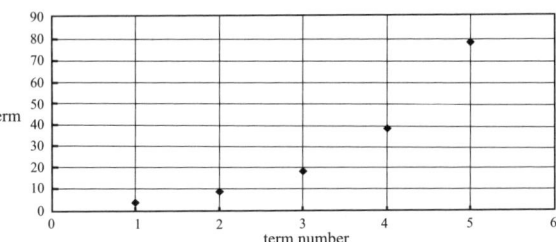

[VCAA 2006 FM]

Question 103

Paula started a stamp collection. She decided to buy a number of new stamps every week. The number of stamps bought in the nth week, t_n, is defined by the difference equation $t_n = t_{n-1} + t_{n-2}$ where $t_1 = 1$ and $t_2 = 2$.
The **total** number of stamps in her collection after five weeks is

A. 8 B. 12 C. 15 D. 19 E. 24

[VCAA 2006 FM]

Question 104

A healthy eating and gym program is designed to help football recruits build body weight over an extended period of time. Roh, a new recruit who initially weighs 73.4 kg, decides to follow the program.
In the first week he gains 400 g in body weight.

B1. Number patterns and applications

In the second week he gains 380 g in body weight.
In the third week he gains 361 g in body weight.
If Roh continues to follow this program indefinitely, and this pattern of weight gain remains the same, his eventual body weight will be closest to

A. 74.5 kg   **B.** 77.1kg   **C.** 77.3 kg   **D.** 80.0 kg   **E.** 81.4 kg

[VCAA 2006 FM]

Question 105

For the geometric sequence 24, 6, 1.5... the common ratio of the sequence is

A. −18   **B.** 0.25   **C.** 0.5   **D.** 4   **E.** 18

[VCAA 2007 FM]

Question 106

The yearly membership of a club follows an arithmetic sequence. In the club's first year it had 15 members. In its third year it had 29 members. How many members will the club have in the fourth year?

A. 8   **B.** 22   **C.** 36   **D.** 43   **E.** 57

[VCAA 2007 FM]

Question 107

The difference equation
$$t_{n+1} = at_n + 6 \text{ where } t_1 = 5$$
generates the sequence 5, 21, 69, 213... The value of a is

A. −1   **B.** 3   **C.** 4   **D.** 15   **E.** 16

[VCAA 2007 FM]

The following information relates to Questions 108 and 109.

The number of waterfowl living in a wetlands area has decreased by 4% each year since 2003. At the start of 2003 the number of waterfowl was 680.

Question 108

If this percentage decrease continues at the same rate, the number of waterfowl in the wetlands area at the start of 2008 will be closest to

A. 532   **B.** 544   **C.** 554   **D.** 571   **E.** 578

[VCAA 2007 FM]

B1. Number patterns and applications

Question 109

W_n is the number of waterfowl at the start of the nth year. Let $W_1 = 680$.
The rule for a difference equation that can be used to model the number of waterfowl in the wetlands area over time is

A. $W_{n+1} = W_n - 0.04n$ **B.** $W_{n+1} = 1.04 W_n$ **C.** $W_{n+1} = 0.04 W_n$
D. $W_{n+1} = -0.04 W_n$ **E.** $W_{n+1} = 0.96 W_n$

[VCAA 2007 FM]

Question 110

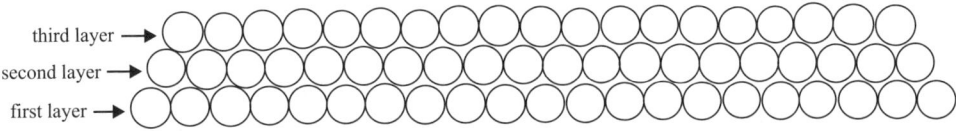

In the first three layers of a stack of soup cans there are 20 cans in the first layer, 19 cans in the second layer and 18 cans in the third layer.
This pattern of stacking cans in layers continues. The maximum number of cans that can be stacked in this way is

A. 190   **B.** 210   **C.** 220   **D.** 380   **E.** 590

[VCAA 2007 FM]

Question 111

The first term, t_1, of a geometric sequence is positive. The common ratio of this sequence is negative. A graph that could represent the first five terms of this sequence is

A.
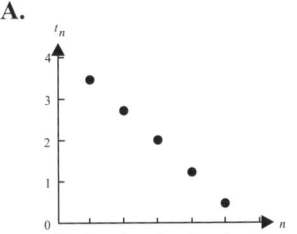

B.
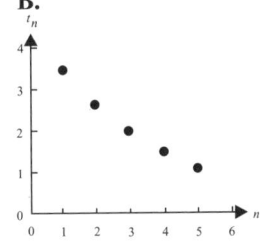

C.
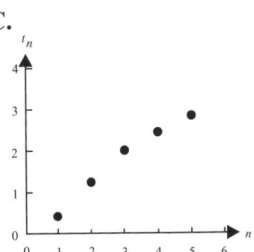

D.
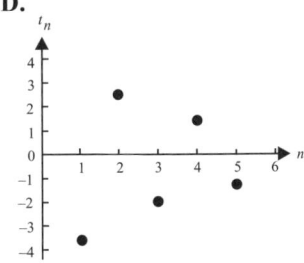

E.
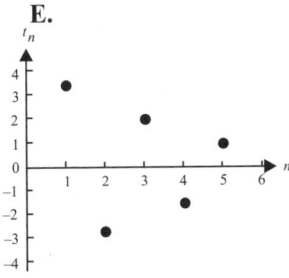

[VCAA 2007 FM]

33

B1. Number patterns and applications

Question 112

The first four terms of a sequence are 12, 18, 30, 54. A difference equation that generates this sequence is

A. $t_{n+1} = t_n + 6$ $t_1 = 12$
B. $t_{n+1} = 1.5 t_n$ $t_1 = 12$
C. $t_{n+1} = 0.5 t_n + 12$ $t_1 = 12$
D. $t_{n+1} = 2 t_n - 6$ $t_1 = 12$
E. $t_{n+2} = t_{n+1} + t_n$ $t_1 = 12, \; t_2 = 18$

[VCAA 2007 FM]

Question 113

At the end of the first day of a volcanic eruption, 15 km² of forest was destroyed. At the end of the second day, an additional 13.5 km² of forest was destroyed. At the end of the third day, an additional 12.15 km² of forest was destroyed. The total area of the forest destroyed by the volcanic eruption continues to increase in this way. In square kilometres, the total amount of forest destroyed by the volcanic eruption at the end of the fourteenth day is closest to

A. 116   B. 119   C. 150   D. 179   E. 210

[VCAA 2007 FM]

B2. Geometry and trigonometry

Question 114

For the right-angled triangle ABC, with BC = 16 cm and AC = 25 cm, the size of angle BAC is closest to

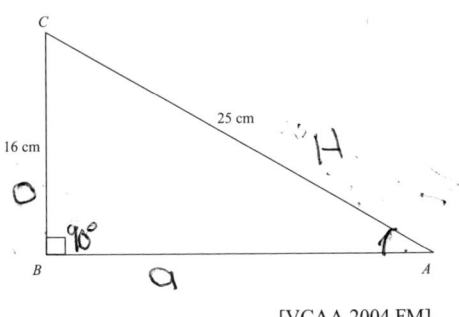

A. 7° B. 25° C. 33°
D. 38° E. 40°

[VCAA 2004 FM]

The following information relates to questions 115 and 116.

The diagram shows the route of a cross-country race. Point X lies due west of point Z.

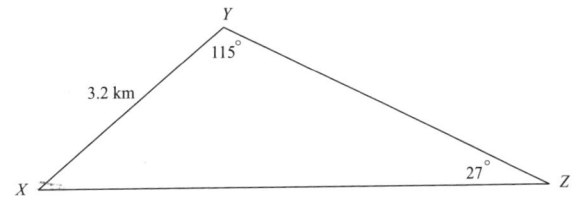

Question 115

The bearing of point Y from point X is

A. 038° B. 052° C. 063° D. 218° E. 232°

[VCAA 2004 FM]

Question 116

Given that the length XY is 3.2 km, the length XZ is closest to

A. 1.5 km B. 1.6 km C. 6.4 km D. 7.0 km E. 7.6 km

[VCAA 2004 FM]

Question 117

A triangle has sides of length 20 cm, 48 cm and 52 cm. A second triangle which is similar to the first triangle has a longest side of 65 cm. The perimeter of the second triangle is

A. 96 cm B. 120 cm C. 125 cm D. 133 cm E. 150 cm

[VCAA 2004 FM]

35

B2. Geometry and trigonometry

Question 118

The points U and V lie on the contour map as shown. The horizontal distance between the two points U and V is 200 metres. The average slope along the line UV is

A. 0.75
B. 1.33
C. 1.50
D. 2.25
E. 2.50

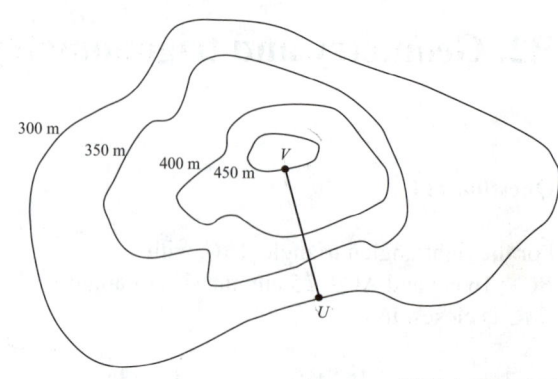

[VCAA 2004 FM]

Question 119

In the diagram, the length of FH is equal to

A. $4 \tan 55°$
B. $\sqrt{(12^2 - 9^2)}$
C. $\sqrt{(12^2 + 9^2 - 216\cos 35°)}$
D. $\dfrac{4\sin 20°}{\sin 15°}$
E. $12\sin 55°$

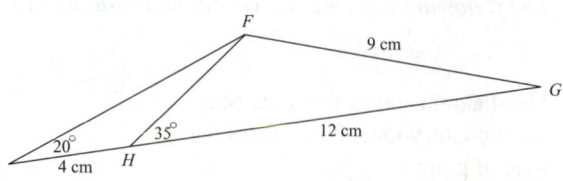

[VCAA 2004 FM]

Question 120

A right-triangular prism $ABCDEF$ is as shown with lengths $AB = 20$ cm, $AC = 100$ cm and $CD = 50$ cm. The size of angle ADB is

A. 10.1°
B. 11.3°
C. 30.5°
D. 59.0°
E. 59.5°

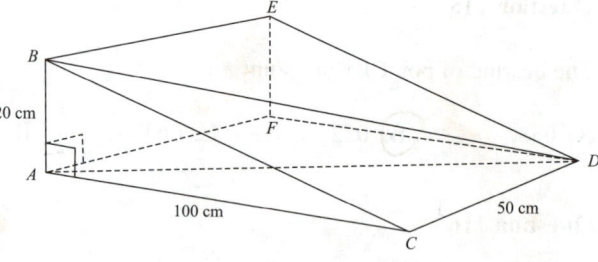

[VCAA 2004 FM]

Question 121

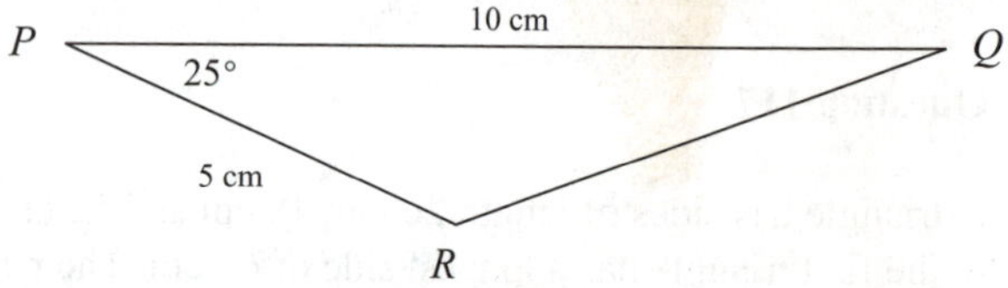

The area of the triangle PQR is closest to

A. 3.3 cm².
B. 6.3 cm².
C. 10.6 cm².
D. 12.5 cm².
E. 22.7 cm².

[VCAA 2005 FM]

B2. Geometry and trigonometry

Question 122

ABCDEF is a regular hexagon. The size of obtuse angle FOD is

A. 60°
B. 90°
C. 120°
D. 135°
E. 165°

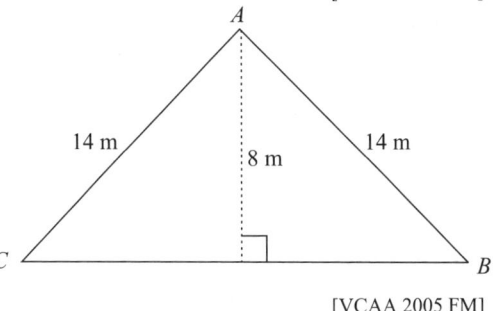

[VCAA 2005 FM]

Question 123

In triangle ABC, the length CB is closest to

A. 11.5 m
B. 12.0 m
C. 16.1 m
D. 19.8 m
E. 23.0 m

[VCAA 2005 FM]

Question 124

Two ships are observed from point O. At a particular time their positions A and B are as shown. The distance between the ships at this time is

A. 3.0 km   B. 3.2 km
C. 4.5 km   D. 9.7 km
E. 10.4 km

[VCAA 2005 FM]

Question 125

The bearing of an aeroplane, X, from a control tower, T, is 055°. Another aeroplane, Y, is due east of control tower T. The bearing of aeroplane X from aeroplane Y is 302°. The size of the angle TXY is

A. 32°   B. 35°   C. 55°   D. 58°   E. 113°

[VCAA 2005 FM]

B2. Geometry and trigonometry

Question 126

In triangle ABC, sin α° = 0.8, sin β° is equal to

A. 0.5
B. 0.6 *(circled)*
C. 0.75
D. 0.8
E. 0.9375

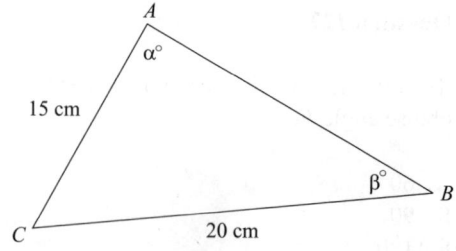

[VCAA 2005 FM]

Question 127

The cube shown has sides four metres long. M is the midpoint of DC. The angle EMH is closest to

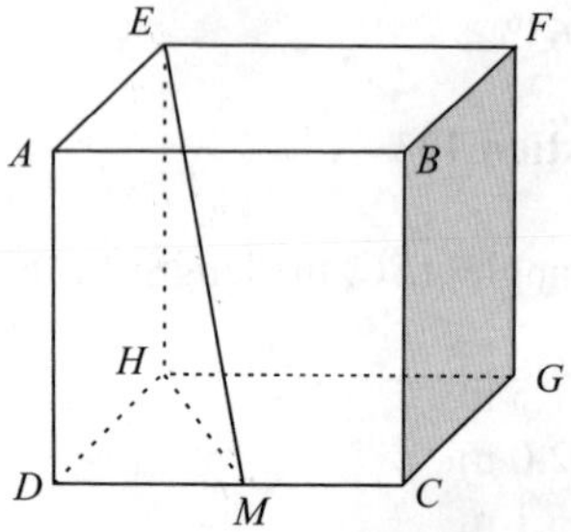

A. 41.8°
B. 48.2°
C. 49.1°
D. 54.7°
E. 70.5°

[VCAA 2005 FM]

Question 128

For the triangle shown, the size of angle θ is closest to

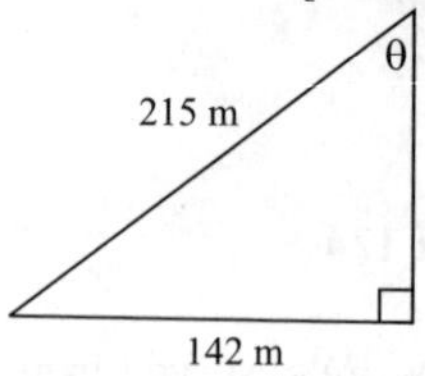

A. 33°
B. 41° *(circled)*
C. 45°
D. 49°
E. 57°

[VCAA 2006 FM]

Question 129

The length of RT in the triangle shown is closest to

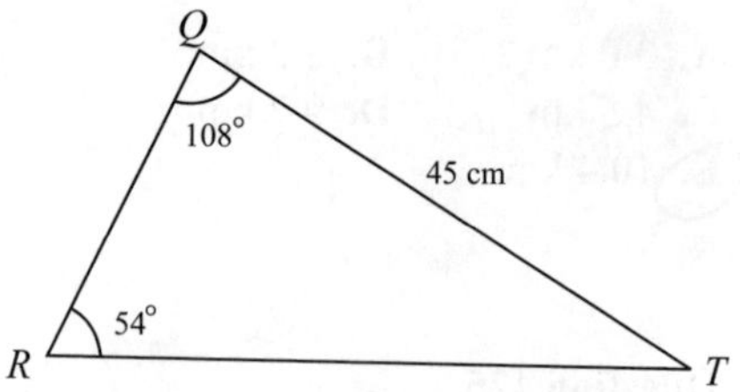

A. 17cm
B. 33cm
C. 45cm
D. 53cm *(circled)*
E. 57cm

[VCAA 2006 FM]

Question 130

The contour map uses 50-metre intervals. The difference in height between the locations represented by points A and B is

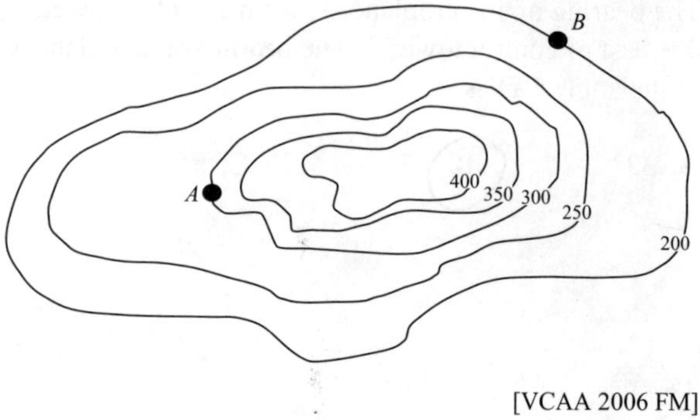

A. 100m *(circled)*
B. 150m
C. 200m
D. 250m
E. 300m

[VCAA 2006 FM]

B2. Geometry and trigonometry

Question 131

The building shown in the diagram is 8 m wide and 24 m long. The side walls are 4 m high. The peak of the roof is 6 m vertically above the ground. In cubic metres, the volume of this building is

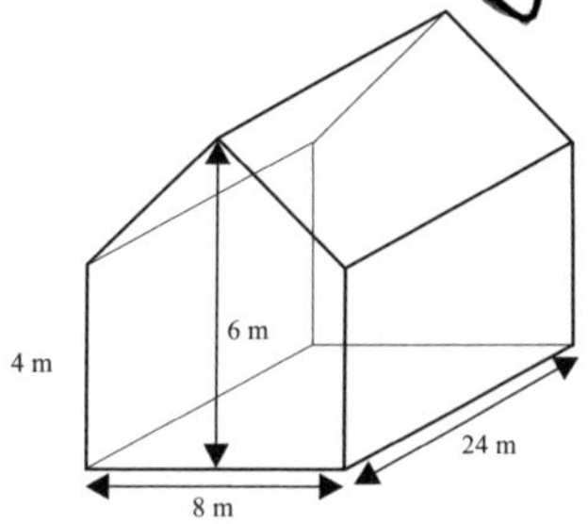

A. 384   B. 576   C. 960
D. 1152   E. 4608

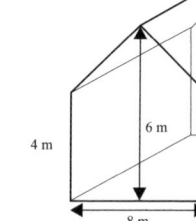

[VCAA 2006 FM]

Question 132

A block of land is triangular in shape. The three sides measure 36 m, 58 m and 42 m. To calculate the area, Heron's formula is used. The correct application of Heron's formula for this triangle is

A. Area = $\sqrt{136(136-36)(136-58)(136-42)}$
B. Area = $\sqrt{136(136-18)(136-29)(136-21)}$
C. Area = $\sqrt{68(68-36)(68-58)(68-42)}$
D. Area = $\sqrt{68(68-18)(68-29)(68-21)}$
E. Area = $\sqrt{68(136-36)(136-58)(136-42)}$

[VCAA 2006 FM]

Question 133

The rectangular box shown in this diagram is closed at the top and at the bottom.
It has a volume of 6 m³. The base dimensions are 1.5 m × 2 m. The total surface area of this box is

A. 10 m²   B. 13 m²
C. 13.5 m² D. 20 m²
E. 27 m²

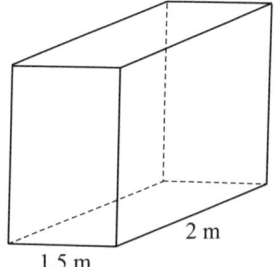

[VCAA 2006 FM]

Question 134

In the diagram, $AD = 9$ cm, $AC = 24$ cm and $DB = 27$ cm. Line segments AC and DE are parallel. The length of DE is

A. 6cm   B. 8cm
C. 12cm   D. 16cm
E. 18cm

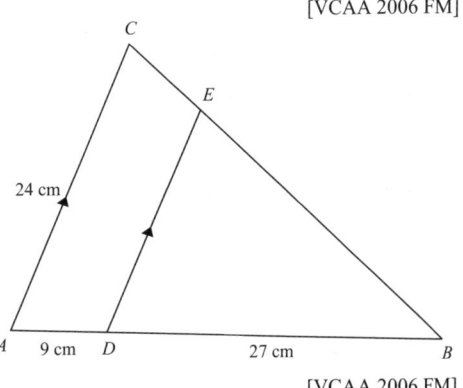

[VCAA 2006 FM]

B2. Geometry and trigonometry

Question 135

The cross-section of a water pipe is circular with a radius, r, of 50 cm, as shown here. The surface of the water has a width, w, of 80 cm. The depth of water in the pipe, d, could be

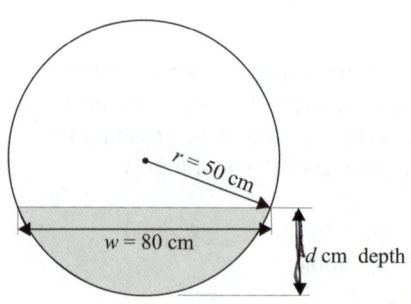

A. 20cm B. 25cm
C. 30cm D. 40cm
E. 50cm

[VCAA 2006 FM]

Question 136

Points M and P are the same distance from a third point O. The bearing of M from O is 038° and the bearing of P from O is 152°. The bearing of P from M is

A. between 000° and 090°
B. between 090° and 180°
C. exactly 180°
D. between 180° and 270°
E. between 270° and 360°

[VCAA 2006 FM]

Question 137

For the triangle shown, the value of $\cos \theta°$ is equal to

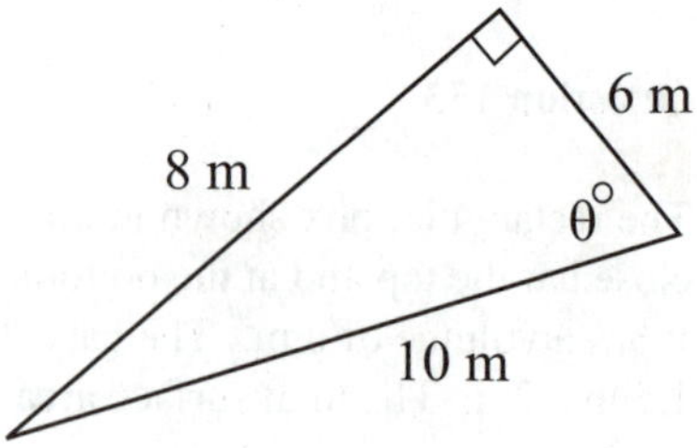

A. $\dfrac{6}{10}$ B. $\dfrac{6}{8}$ C. $\dfrac{8}{10}$

D. $\dfrac{10}{8}$ E. $\dfrac{8}{6}$

[VCAA 2007 FM]

Question 138

For an observer on the ground at A, the angle of elevation of a weather balloon at B is 37°. C is a point on the ground directly under the balloon.
The distance AC is 2200 m.
To the nearest metre, the height of the weather balloon above the ground is

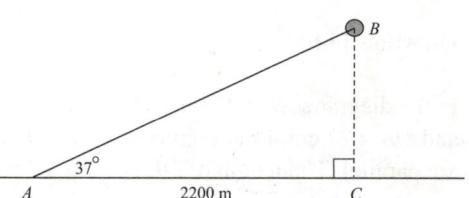

A. 1324 m   B. 1658 m   C. 1757m   D. 2919 m   E. 3655 m

[VCAA 2007 FM]

B2. Geometry and trigonometry

Question 139

A rectangle is 3.79 m wide and has a perimeter of 24.50 m. Correct to one decimal place, the length of the diagonal of this rectangle is

A. 9.2 m   B. 9.3 m   C. 12.2 m   D. 12.3 m   E. 12.5 m

[VCAA 2007 FM]

Question 140

A steel beam used for constructing a building has a cross-sectional area of 0.048 m^2 as shown. The beam is 12 m long.
In cubic metres, the volume of this steel beam is closest to

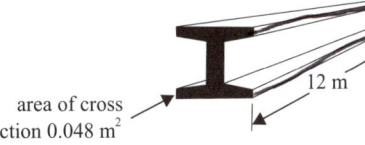
area of cross section 0.048 m^2 12 m

A. 0.576   B. 2.5   C. 2.63   D. 57.6   E. 2500

[VCAA 2007 FM]

Question 141

A block of land has an area of 4000 m^2. When represented on a map, this block of land has an area of 10 cm^2. On the map 1 cm would represent an actual distance of

A. 10 m   B. 20 m   C. 40 m   D. 400 m   E. 4000 m

[VCAA 2007 FM]

Question 142

A solid cylinder has a height of 30 cm and a diameter of 40 cm. A hemisphere is cut out of the top of the cylinder as shown here. In square centimetres, the total surface area of the remaining solid (including its base) is closest to

30 cm

40 cm

A. 1260   B. 2510   C. 6280   D. 7540   E. 10 050

[VCAA 2007 FM]

Question 143

A closed cubic box of side length 36 cm is to contain a thin straight metal rod. The maximum possible length of the rod is closest to

A. 36 cm   B. 51 cm   C. 62 cm   D. 108 cm   E. 216cm

[VCAA 2007 FM]

B2. Geometry and trigonometry

Question 144

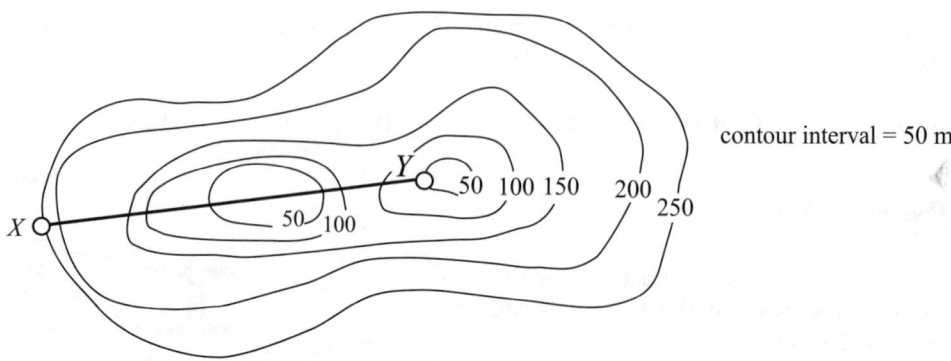

For the contour map shown here, the cross section along the line segment XY could be

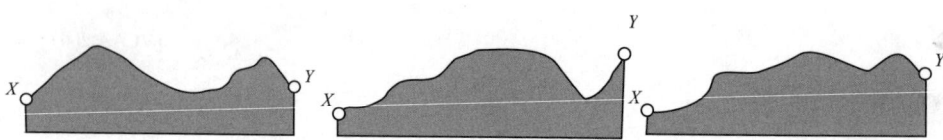

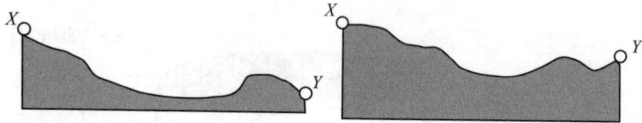

[VCAA 2007 FM]

Question 145

The points M, N and P form the vertices of a triangular course for a yacht race.
$MN = MP = 4$ km.
The bearing of N from M is 070° and the bearing of P from M is 180°
Three people perform different calculations to determine the length of NP in kilometres.

Graeme $NP = \sqrt{16 + 16 - 2 \times 4 \times 4 \times \cos 110°}$
Shelley $NP = 2 \times 4 \times \cos 35°$
Tran $NP = \dfrac{4 \times \sin 110°}{\sin 35°}$ The correct length of NP would be found by

A. Graeme only. B. Tran only. C. Graeme and Shelley only.
D. Graeme and Tran only.
E. Graeme, Shelley and Tran.

[VCAA 2007 FM]

B3. Graphs and relations

Question 146

The graph shows the cost (dollars) of mobile telephone calls up to 240 seconds long.

The cost of making a 90-second call followed by a 30-second call is

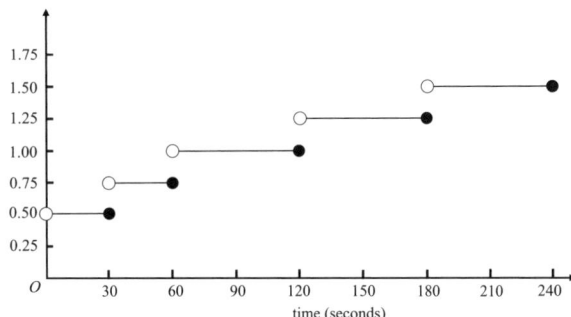

A. $1.00 **B.** $1.20 **C.** $1.25 **D.** $1.50 **E.** $1.75

[VCAA 2004 FM]

Question 147

The point (2, 1) lies on the line $y = 3x + c$. The value of c is

A. −7 **B.** −5 **C.** −1 **D.** 5 **E.** 7

[VCAA 2004 FM]

Question 148

The lines $y + 8 = 0$ and $x − 12 = 0$ intersect at the point

A. (−12, 8) **B.** (−8, 12) **C.** (0, 0) **D.** (8, −12) **E.** (12, −8)

[VCAA 2004 FM]

The following information relates to questions 149 and 150.

The graph shows a distance-time graph for a car travelling from home along a long straight road over a 16-hour period.

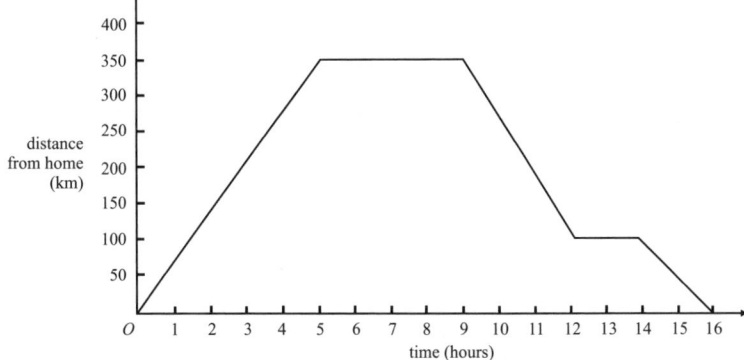

43

B3. Graphs and relations

Question 149

In which one of the time intervals is the speed of the car greatest?

A. 0 to 5 hours
B. 5 to 9 hours
C. 9 to 12 hours
D. 12 to 14 hours
E. 14 to 16 hours

[VCAA 2004 FM]

Question 150

After twelve hours the car has travelled a total distance of

A. 100km   B. 350km   C. 450km   D. 600km   E. 700km

[VCAA 2004 FM]

Question 151

The cost, $C, of hiring a boat for x hours is given by the equation $C = ax + b$ where a is the hourly rate and b is a fixed booking fee. When the boat is hired for 4 hours the cost is $320. When the boat is hired for 6 hours the cost is $450. When the boat is hired for one hour the cost is

A. $65   B. $75   C. $77   D. $80   E. $125

[VCAA 2004 FM]

Question 152

The shaded region shown in the graph (with boundaries included) represents the feasible region for a linear programming problem. The maximum value of the objective function $y - 2x + 20$, for this feasible region, is

A. 18
B. 23
C. 25
D. 27
E. 33

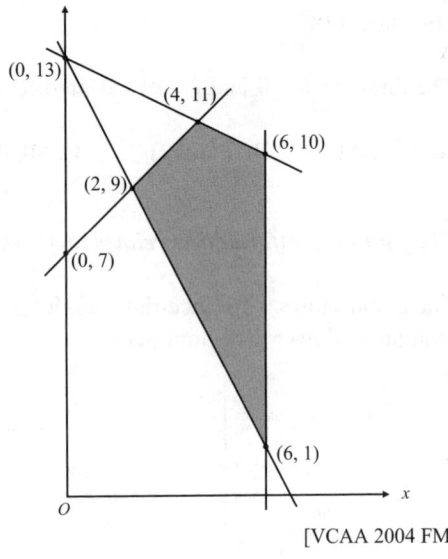

[VCAA 2004 FM]

B3. Graphs and relations

Question 153

The shaded region shown in the graph (with boundaries included) is described by

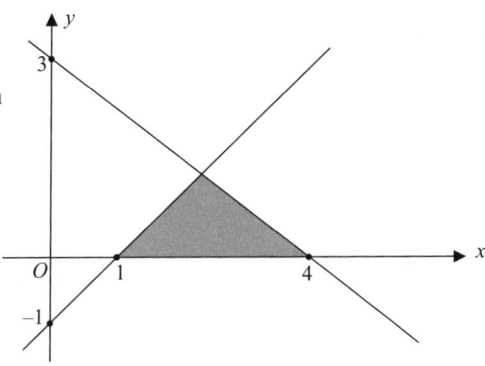

A. $3x + 4y \leq 12$
$x - y \leq 1$
$x \geq 0$
$x \geq 0$

B. $3x + 4y \leq 12$
$x - y \geq 1$
$x \geq 0$
$x \geq 0$

C. $3x + 4y \geq 12$
$x - y \geq 1$
$x \geq 0$
$x \geq 0$

D. $4x + 3y \leq 12$
$x - y \leq 1$
$x \geq 0$
$x \geq 0$

E. $4x + 3y \leq 12$
$x - y \leq 1$
$x \geq 0$
$x \geq 0$

[VCAA 2004 FM]

Question 154

The graph shown represents a relationship $y = kx^3$. A graph that shows this relationship when y is plotted against x is

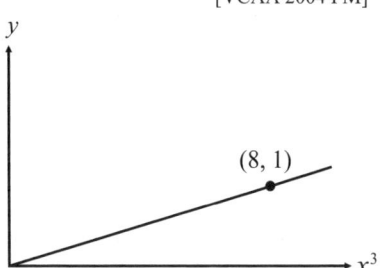

A.
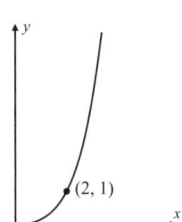
(2, 1)

B.
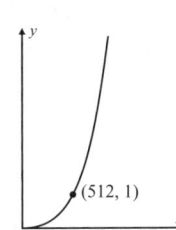
(512, 1)

C.
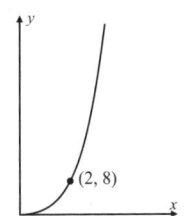
(2, 8)

D.
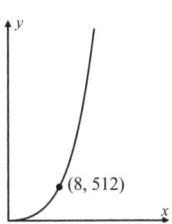
(8, 512)

E.
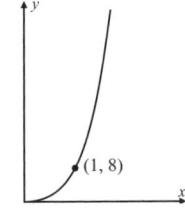
(1, 8)

[VCAA 2004 FM]

B3. Graphs and relations

Question 155

The graph shows the temperature (in degrees Celsius) over a 24-hour period.

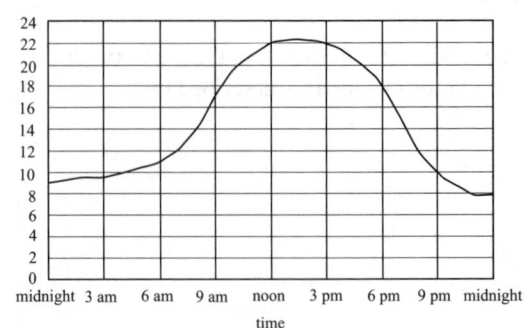

The period of greatest temperature increase was

A. 6 am – 9 am
B. 9 am – noon
C. noon – 3 pm
D. 3pm – 6pm
E. 6pm – 9pm

[VCAA 2005 FM]

Question 156

Two lines have equations $y = -5$ and $y = -x + 5$ respectively. The point that lies on both of these lines is

A. (–10, 5)   B. (–5, 5)   C. (0, –5)   D. (5, –5)   E. (10, –5)

[VCAA 2005 FM]

Question 157

Which one of the following statements is **not** true?

A. The line with equation $7x - 4y = 0$ passes through the point (4, 7).
B. The point (3, 5) lies in the region defined by $7x - 4y \geq 0$.
C. The line with equation $3x + 5y = 0$ has a positive gradient.
D. The lines $7x - 4y = 0$ and $3x + 5y = 0$ meet at the origin.
E. For the line with the equation $7x - 4y = 0$, y increases as x increases.

[VCAA 2005 FM]

Question 158

If the line shown has equation $3x + 2y = 4k$, then the value of k must be

A. 2
B. 3
C. 6
D. 8
E. 12

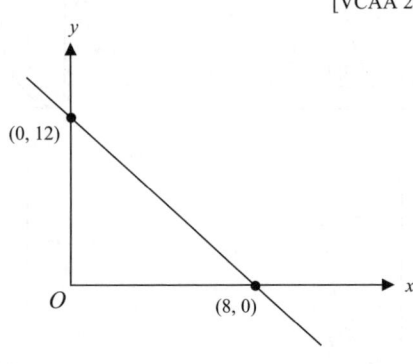

[VCAA 2005 FM]

46

B3. Graphs and relations

Question 159

An electrician charges a fixed call-out fee of $50 and then charges $65 per hour for each hour worked. For n hours worked, the total charge in dollars is

A. 115   B. $n + 115$ C. $50n + 65$ D. $65n + 50$ E. $115n$

[VCAA 2005 FM]

Question 160

One afternoon at the beach Mr Smith bought four ice creams and three drinks for his family at a cost of $21.40. Mrs Brown bought five of the same ice creams and two of the same drinks for $20.80. Based on these prices, the cost of one drink is

A. $2.80   B. $2.90   C. $3.00   D. $3.30   E. $3.40

[VCAA 2005 FM]

Question 161

The relationship between the variables a and b as shown in the graph is

A. $b = a^2$
B. $b = 2a^2$
C. $b = 2a$
D. $a = 2b$
E. $a^2 = 2b$

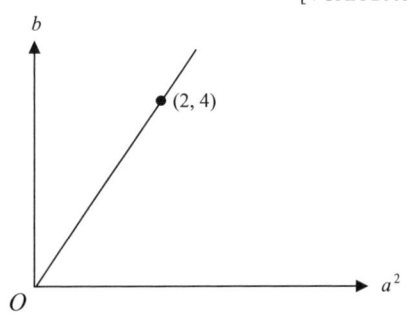

[VCAA 2005 FM]

Question 162

For the shaded region (with boundaries included), the value of the objective function $P = 4x - 3y$ is a **maximum** at the point

A. (0, 0)
B. (0, 100)
C. (50, 100)
D. (90, 60)
E. (120, 0)

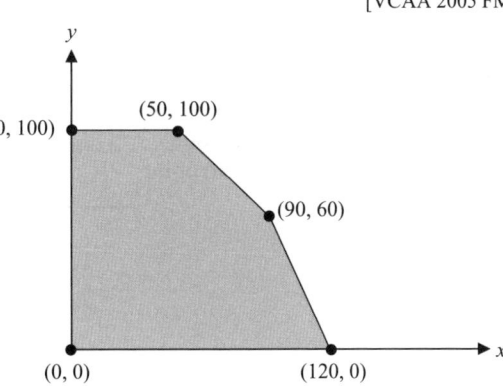

[VCAA 2005 FM]

B3. Graphs and relations

Question 163

The relationship between the intensity of sound, and the distance from the source of the sound, is $I = \dfrac{k}{d^2}$, where I is the intensity of sound measured in watts per square metre (W/m^2), d is the distance in metres from the source, and k is a constant. The intensity of sound is 20 W/m^2 at a distance of 50 m from the source. A graph showing the relationship between intensity of sound and distance from the source is

A.

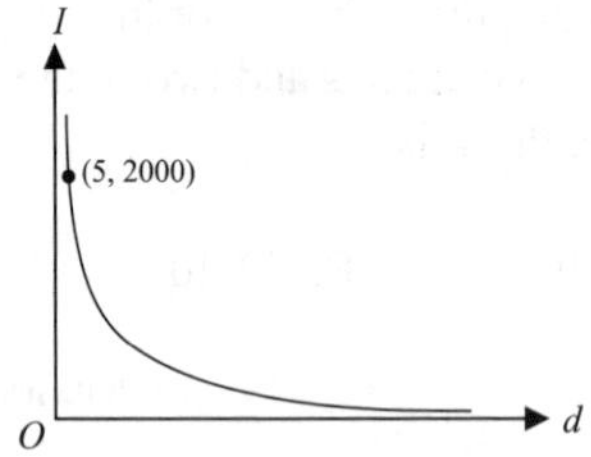

B.

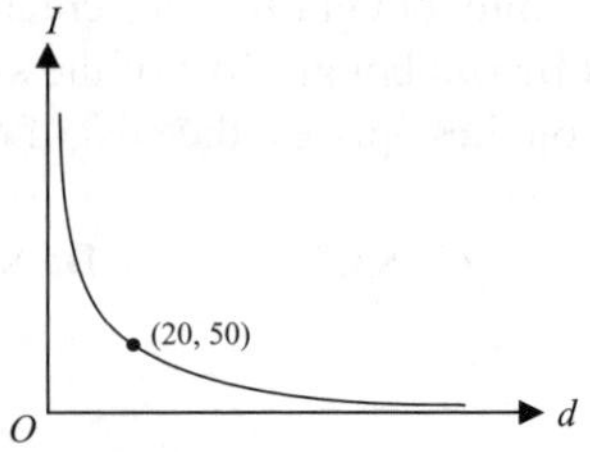

C.

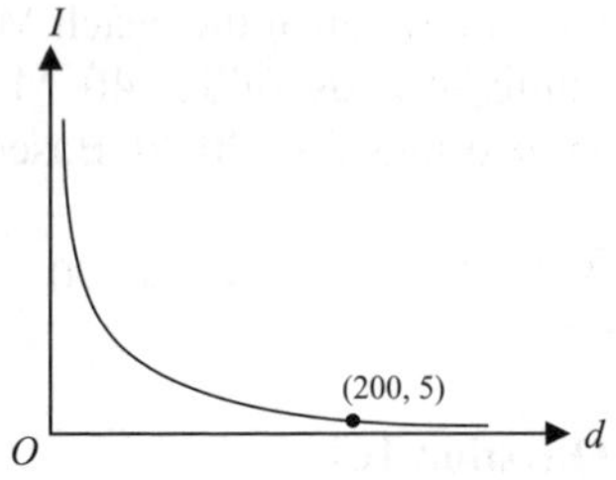

D.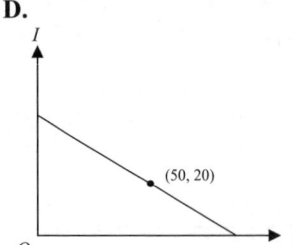

E.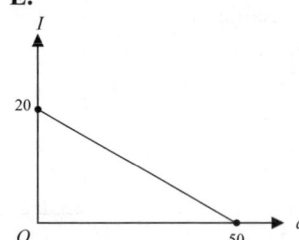

[VCAA 2005 FM]

Question 164

On the graph shown, the equation of the line passing through the point (1, 2) is

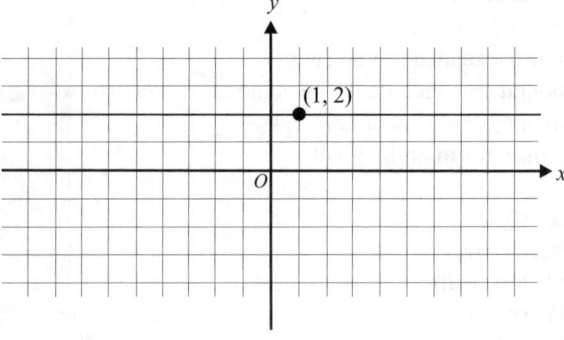

A. $x = 1$
B. $y = 1$
C. $x = 2$
D. $y = 2$
E. $y = x + 1$

[VCAA 2006 FM]

48

B3. Graphs and relations

Question 165

The graph shown represents the depth of water in a channel (in metres) as it changes over time (in hours). During the time interval shown, the number of times the depth of the water in the channel is 10 metres is

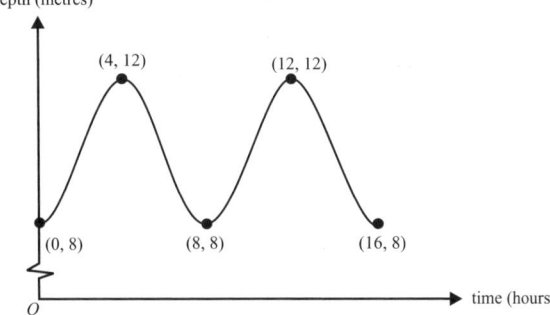

A. 0
B. 1
C. 2   D. 3   E. 4

[VCAA 2006 FM]

The following graph relates to Questions 166 and 167.

A gas-powered camping lamp is lit and the gas is left on for six hours. During this time the lamp runs out of gas. The graph shows how the mass, M, of the gas container (in grams) changes with time, t (in hours), over this period.

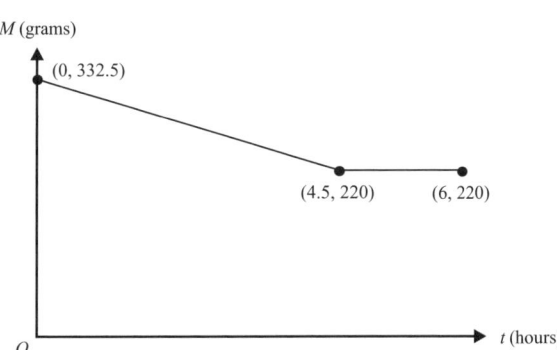

Question 166

Assume that the loss in weight of the gas container is due only to the gas being burnt. From the graph it can be seen that the lamp runs out of gas after

A. 1.5 hours.   B. 3 hours.   C. 4.5 hours.   D. 6 hours.   E. 220 hours.

[VCAA 2006 FM]

Question 167

Which one of the following rules could be used to describe the graph above?

A. $M = \begin{cases} 332.5 - 25t & \text{for } 0 \leq t \leq 4.5 \\ 220 & \text{for } 4.5 < t \leq 6 \end{cases}$

B. $M = \begin{cases} 332.5 - 25t & \text{for } 0 \leq t \leq 4.5 \\ 220t & \text{for } 4.5 < t \leq 6 \end{cases}$

C. $M = \begin{cases} 332.5 + 25t & \text{for } 0 \leq t \leq 4.5 \\ 220t & \text{for } 4.5 < t \leq 6 \end{cases}$

B3. Graphs and relations

D. $M = \begin{cases} 332.5 - 12.5t & \text{for } 0 \le t \le 4.5 \\ 220t & \text{for } 4.5 < t \le 6 \end{cases}$

E. $M = \begin{cases} 332.5 - 12.5t & \text{for } 0 \le t \le 4.5 \\ 220 & \text{for } 4.5 < t \le 6 \end{cases}$

[VCAA 2006 FM]

Question 168

Which one of the following statements about the line with equation $12x - 4y = 0$ is **not** true?

A. the line passes through the origin
B. the line has a slope of 12
C. the line has the same slope as the line with the equation $12x - 4y = 12$
D. the point (1, 3) lies on the line
E. for this line, as x increases y increases

[VCAA 2006 FM]

Question 169

The point of intersection of two lines is $(2, -2)$. One of these two lines could be

A. $x - y = 0$
B. $2x + 2y = 8$
C. $2x + 2y = 0$
D. $2x - 2y = 4$
E. $2x - 2y = 0$

[VCAA 2006 FM]

Question 170

In a linear programming problem involving animal management on a farm

- x represents the number of cows on the farm
- y represents the number of sheep on the farm.

The feasible region (with boundaries included) for the problem is indicated by the shaded region on the diagram here.

One of the constraints defining the feasible region indicates that

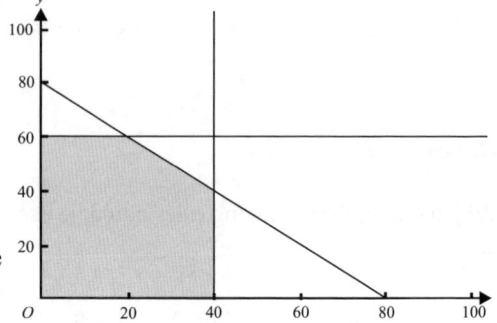

A. there must be 20 cows and 60 sheep.
B. there must be 40 cows and 40 sheep.
C. the number of sheep cannot exceed 40.
D. the number of cows must be at least 60.
E. the total number of cows and sheep cannot exceed 80.

[VCAA 2006 FM]

B3. Graphs and relations

Question 171

The cost of manufacturing a number of frying pans consists of a fixed cost of $400 plus a cost of $50 per frying pan. The manufacturer could break even by selling

- A. 10 frying pans at $90 each.
- B. 10 frying pans at $45 each.
- C. 15 frying pans at $60 each.
- D. 15 frying pans at $30 each.
- E. 20 frying pans at $50 each.

[VCAA 2006 FM]

Question 172

The four inequalities below were used to construct the feasible region for a linear programming problem.

$$x \geq 0$$
$$y \geq 0$$
$$x + y \leq 9$$
$$y \leq \frac{1}{2}x$$

A point that lies within this feasible region is

A. (4, 4)   B. (5, 3)   C. (6, 2)   D. (6, 4)   E. (7, 3)

[VCAA 2006 FM]

Question 173

The line shown passes through the origin and the point (2, 1). The slope of this line is

A. −2   B. −1   C. $-\frac{1}{2}$ D. $\frac{1}{2}$ E. 2

[VCAA 2007 FM]

Question 174

A builder's fee, C dollars, can be determined from the rule $C = 60 + 55n$, where n represents the number of hours worked.
According to this rule, the builder's fee will be

- A. $60 for 1 hour of work.
- B. $110 for 2 hours of work.
- C. $500 for 8 hours of work.
- D. $550 for 10 hours of work.
- E. $1150 for 10 hours of work.

[VCAA 2007 FM]

B3. Graphs and relations

Question 175

The graph shown represents the temperature. in degrees Celsius, over a 16-hour period. During this period, the minimum temperature occurred at

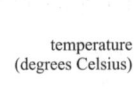
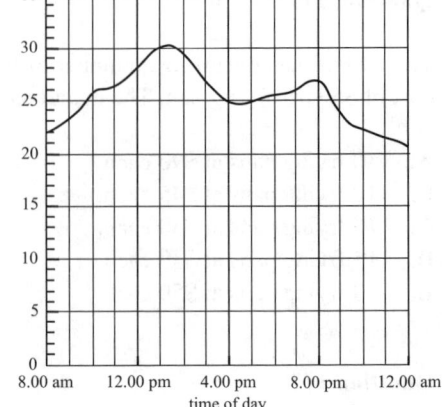

A. 8.00 am
B. 1.20 pm
C. 4.15 pm
D. 7.45 pm
E. 12.00 am

[VCAA 2007 FM]

Question 176

Paul makes rulers. There is a fixed cost of $60 plus a manufacturing cost of $0.20 per ruler. Last week Paul was able to break even by selling his rulers for $1 each. The number of rulers Paul sold last week was

A. 50   B. 75   C. 90   D. 120   E. 150

[VCAA 2007 FM]

Question 177

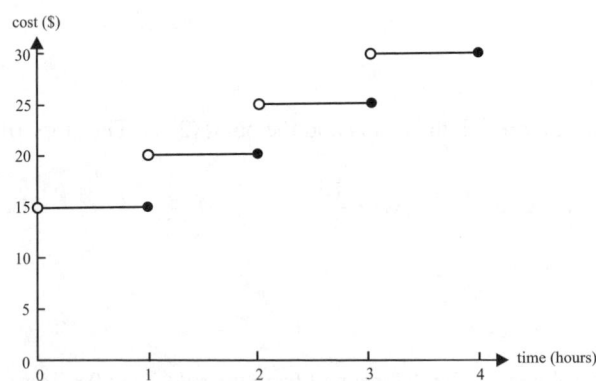

The cost of hiring one motorbike for up to 4 hours is shown in the graph above.
Two motor bikes were hired. The total charge for hiring the two motorbikes was $45.
The time for which each motorbike was hired could have been

A. 1 hour and 2 hours.
B. 1 hour and 3 hours.
C. 1.5 hour and 2 hours.
D. 1.5 hour and 3 hours.
E. 2 hour and 3.5 hours.

[VCAA 2007 FM]

52

B3. Graphs and relations

Question 178

Russell is a wine producer. He makes both red and white wine. Let x represent the number of bottles of red wine he makes and y represent the number of bottles of white wine he makes. This year he plans to make at least twice as many bottles of red wine as white wine. An inequality representing the situation is

A. $y \leq x + 2$
B. $y \leq 2x$
C. $y \geq 2x$
D. $x \leq 2y$
E. $x \geq 2y$

[VCAA 2007 FM]

Question 179

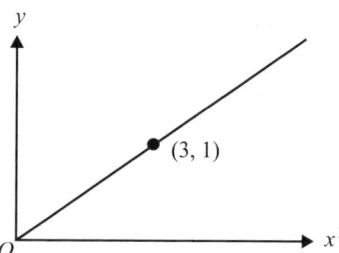

The graph above shows a relationship between y and x^3. The graph that shows the same relationship between y and x is

A.
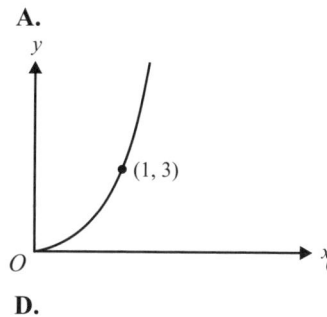

B.
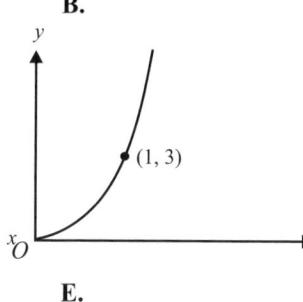

C.
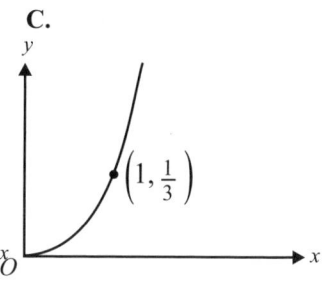

D.
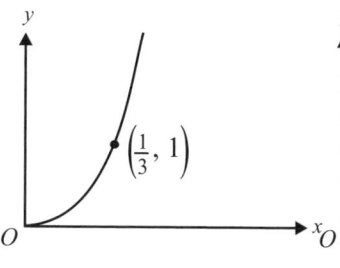

E.
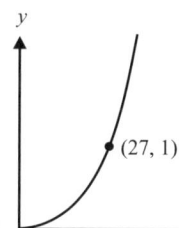

[VCAA 2007 FM]

B3. Graphs and relations

Question 180

Which one of the following pairs of simultaneous linear equations has no solution?

A. $3x - y = 5$
    $4x + y = 9$

B. $2x - y = 1$
    $4x - 2y = 3$

C. $x + 3y = 0$
    $2x - y = 7$

D. $x - 3y = 10$
    $3x + 2y = 8$

E. $4x + y = -6$
    $2x - y = 0$

[VCAA 2007 FM]

Question 181

The following five constraints apply to a linear programming problem.

$x \geq 0,$ $y \geq 0,$ $x + y \geq 50,$
$x + y \leq 100,$ $y \leq x$

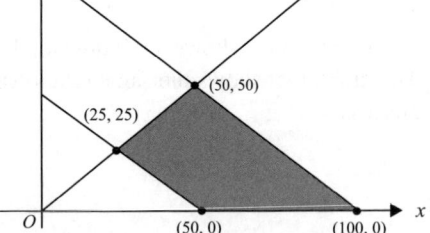

In the diagram, the shaded region (with boundaries included) represents the feasible region for this linear programming problem. The aim is to maximise the objective function $Z = 2x + ky$. If the maximum value of Z occurs only at the point (100, 0), then a possible value for k is

A. 1   B. 2   C. 3   D. 4   E. 5

[VCAA 2007 FM]

B4. Business-related mathematics

Question 182

Under a hire purchase agreement, Sheng will pay a total of $960 for a television set. He is required to pay a deposit of $120 and to pay the balance in regular equal monthly payments over 6 months. The monthly repayments are

A. $70   B. $80   C. $140   D. $160   E. $180

[VCAA 2003 FM]

Question 183

Derek invested $26 000 for eighteen months and earned $975 in simple interest. The annual interest rate for the investment is

A. 0.025%   B. 0.0563%   C. 2.5%   D. 3.75%   E. 5.63%

[VCAA 2003 FM]

Question 184

Heather invests $45 000 at 4% per annum for 5 years compounding annually. The total amount of interest earned is

A. $1800   B. $2100   C. $9000   D. $9750   E. $54 750

[VCAA 2003 FM]

Question 185

Swee borrowed $150 000 at 6.2% per annum compounding monthly. The repayments are $1100 per month. The balance of the loan at the end of five years is closest to

A. $0   B. $84 000   C. $127 000   D. $137 000   E. $148 000

[VCAA 2003 FM]

Question 186

Zoltan is running a convenience store. He purchases equipment for $6500. It is anticipated that the equipment will last 5 years and have a depreciated value of $2000. Assuming the straight line method of depreciation, the equipment is depreciated annually by

A. $400   B. $900   C. $1027   D. $1300   E. $4500

[VCAA 2003 FM]

B4. Business-related mathematics

Question 187

Interest is paid monthly into an account at a rate of 3% per annum. Each month, immediately after the interest is paid, the account is debited $5 in fees. No other transactions take place. The initial amount of money in the account is $12 200. After all the interest has been paid and fees debited, the balance in the account at the end of two months is

A. $12 251.06
B. $12 261.08
C. $12 271.09
D. $12 932.83
E. $12 953.13

[VCAA 2003 FM]

Question 188

Lim invested $8000 in an investment account, earning $r\%$ interest per annum, compounding quarterly. The balance in dollars, after 5 years, is given by

A. $8000\left(1+\dfrac{r}{100}\right)^{5}$
B. $8000\left(1+\dfrac{r}{100}\right)^{20}$
C. $8000\left(1+\dfrac{r}{400}\right)^{5}$

D. $8000\left(1+\dfrac{r}{400}\right)^{20}$
E. $8000\left(1+\dfrac{r}{1200}\right)^{60}$

[VCAA 2003 FM]

Question 189

The following is an extract from a bank account showing all transactions for the period 1 January to 30 June, 2003.

Date	Particulars	Credit	Debit	Balance
01 Jan 2003	Brought Forward			4320.00
15 Mar 2003	Deposit	2100.00		6420.00
31 Mar 2003	Interest	32.40		6452.40
22 May 2003	Withdrawal		460.00	5992.40
30 June 2003	Interest			

Interest on this account is calculated at a rate of 0.25% per month on the minimum monthly balance and paid into the account quarterly. Interest for the June period (April to June) is paid on 30 June. The balance in the account after interest is paid on 30 June 2003 is

A. $6039.64
B. $6038.49
C. $6024.76
D. $6023.51
E. $6022.36

[VCAA 2003 FM]

B4. Business-related mathematics

Question 190

Peter borrows $80 000 for 10 years at 5.6% per annum, compounding monthly, with monthly repayments of $555. Which one of the following statements is true?

A. The loan will be fully paid out in ten years.
B. At the end of five years, the balance of the loan will be $40 000.
C. The amount of interest paid each month during the loan increases.
D. Weekly repayments of $132 compounding weekly would reduce the period of the loan.
E. If one extra payment of $2000 is to be made, it would be better to make it at the end of year eight than at the end of year two.

[VCAA 2003 FM]

Question 191

Sarah invests $37 000 at a simple interest rate of 4% per annum. The total amount of interest earned in two years is

A. $ 1480   B. $2960   C. $5920   D. $38 480   E. $39 960

[VCAA 2004 FM]

Question 192

Ardy invests $150 000 for 6 years at an interest rate of 3.5% per annum, compounding annually. The value of the investment at the end of the 6 years is

A. $31 500.00   B. $34 388.30   C. $178 107.00
D. $181 500.00   E. $184 388.30

[VCAA 2004 FM]

Question 193

Leonard charged $36 per hour for tutoring. In September 2004, he increased his fees by 15%. For two hours of tutoring, Leonard now charges

A. $10.80   B. $41.40   C. $47.60   D. $72.00   E. $82.80

[VCAA 2004 FM]

The following information relates to questions 194 and 195.

Chen buys a new refrigerator. The advertised price was $2700. He chooses to pay a deposit of $500 and monthly repayments of $115 over 2 years.

B4. Business-related mathematics

Question 194

Under this arrangement, the total cost of the refrigerator is

A. $2200   B. $2760   C. $3200   D. $3260   E. $4260

[VCAA 2004 FM]

Question 195

The annual flat rate of interest paid is closest to

A. 10.4%   B. 12.2%   C. 12.7%   D. 20.7%   E. 25.4%

[VCAA 2004 FM]

Question 196

A loan of $250 000 is to be paid back over a period of 20 years at an interest rate of 7.4% per annum, compounding monthly. To the nearest dollar, the monthly repayment is closest to

A. $1963   B. $1999   C. $2998   D. $4343   E. $13 326

[VCAA 2004 FM]

Question 197

Binnie invests $12 000 for 5 years at an interest rate of 3.6% per annum, compounding annually. The amount of interest she earns during the third year of the investment is closest to

A. $463.66   B. $470.41   C. $480.36   D. $1343.22   E. $1823.57

[VCAA 2004 FM]

Question 198

A machine is purchased for $15 000. Using the reducing balance method of depreciation, its book value after six years will be $5000. The graph that best represents the value of the machine at the end of each year over the six-year period is

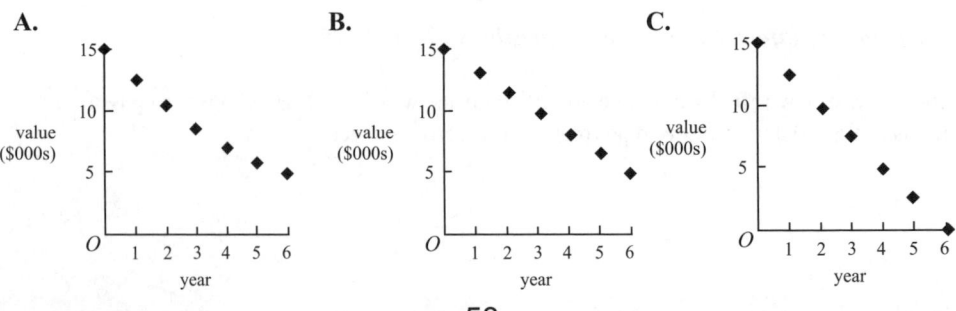

B4. Business-related mathematics

D. 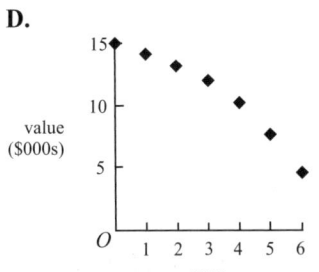   **E.**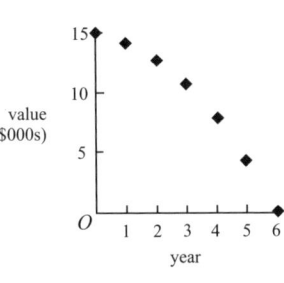

[VCAA 2004 FM]

Question 199

An amount of $130 000 is borrowed at an interest rate of 7.5% per annum, compounding monthly. The loan is fully repaid over ten years with equal monthly repayments. Which of the following statements is **not** true?

- **A.** The monthly interest rate is 0.625%.
- **B.** No money will be owed after 10 years.
- **C.** The total number of repayments is 120.
- **D.** A monthly repayment of $1500 will reduce the length of the loan.
- **E.** At the end of five years, the amount of the principal still owing will exceed $65 000.

[VCAA 2004 FM]

Question 200

An investment of $1200 for one year returns $140.38 interest. The simple interest rate per annum for this investment is closest to

- **A.** 0.88%   **B.** 3.21%   **C.** 4.38%   **D.** 11.70%   **E.** 15.99%

[VCAA 2005 FM]

Question 201

The price of a book is $22. It includes a Goods and Services Tax (GST) of 10%. The price before this Goods and Services Tax (GST) is added is

- **A.** $19.80   **B.** $20.00   **C.** $24.00   **D.** $24.20   **E.** $24.40

[VCAA 2005 FM]

B4. Business-related mathematics

Question 202

Tamara's bank statement for September has been damaged by spilt ink as shown below. Tamara's Pay/Salary was deposited on 30 September. What is the value of this deposit?

Date	Transaction details	Withdrawals ($)	Deposits ($)	Balance ($)
01 Sept	Opening balance			2143.50
01 Sept	Interest		2.45	
13 Sept	Payment - Telstra	616.40		
30 Sept	Pay/Salary			
Totals at the end of month		**616.40**		**1971.75**

A. $441.95 **B.** $442.20 **C.** $444.65 **D.** $785.70 **E.** $788.15

[VCAA 2005 FM]

Question 203

A machine that makes boxes costs $45 000. Its value depreciates by five cents for every box it makes. Each year it makes 120 000 boxes. The depreciated value of this machine at the end of two years is

A. $33 000 **B.** $38 000 **C.** $39 000 **D.** $45 000 **E.** $115 000

[VCAA 2005 FM]

Question 204

An investor borrows $200 000 for five years to buy an apartment. The interest rate is 8.5% per annum compounding monthly. It is an interest only loan, that is, at the end of five years, the investor will still owe $200 000. He is required to make monthly repayments. Correct to the nearest cent, his monthly repayment will be

A. $666.67 **B.** $1416.67 **C.** $1757.67 **D.** $4103.31 **E.** $6789.95

[VCAA 2005 FM]

Question 205

Tim invests $3000 in a term deposit account that adds 6.5% interest annually, calculated on the account balance at the end of each year. The interest paid in the fourth year is

A. $195.00 **B.** $221.16 **C.** $235.55 **D.** $3623.85 **E.** $3859.40

[VCAA 2005 FM]

B4. Business-related mathematics

Question 206

Gregor invests $10000 and earns interest at a rate of 6% per annum compounding quarterly. Every quarter, after interest has been added, he withdraws $500. At the end of four years, after interest has been added and he has made the $500 withdrawal, the value of the remaining investment will be closest to

A. $3720   B. $4220   C. $5440   D. $21 660   E. $22 160

[VCAA 2005 FM]

Question 207

On Monday, a clock has a certain price. In a crazy week of sales the price goes up and down as follows:
Tuesday - the price of the clock is increased by 10%.
Wednesday - Tuesday's price is reduced by 10%.
Thursday - Wednesday's price is increased by 20%.
Friday - Thursday's price is reduced by 20%.
When compared with Monday's price, the price on Friday is closest to

A. the same as Monday's price.
B. 4% lower than Monday's price.
C. 4% higher than Monday's price.
D. 5% lower than Monday's price.
E. 5% higher than Monday's price.

[VCAA 2005 FM]

Question 208

Sally planned to repay a loan fully with six equal monthly repayments of $800. Interest was calculated monthly on the reducing balance. Sally missed the third payment, but made a double payment of $1600 in the fourth month. Which of the following statements is true?

A. The same amount of interest is paid each month.
B. The amount owing after three months is the same as the amount owed after two months.
C. The amount owing after three months is less than the amount owed after two months.
D. To fully repay the loan, Sally will pay less than $4800.
E. To fully repay the loan, Sally will pay more than $4800.

[VCAA 2005 FM]

Question 209

$4000 is invested at a simple interest rate of 5% per annum. The amount of interest earned in the first year is

A. $20   B. $200   C. $220   D. $420   E. $2000

[VCAA 2006 FM]

B4. Business-related mathematics

Question 210

A bank statement for the month of October is shown below.

Date	Description of transaction	Debit	Credit	Balance
01 Oct	Opening balance			853.92
01 Oct	Withdrawal — Internet banking	380.00		473.92
16 Oct	Deposit— Cheque		518.15	992.07
18 Oct	Credit card payment	125.56		866.51
23 Oct	Withdrawal — Internet banking	250.00		616.51
31 Oct	Closing balance			616.51

Interest on this account is calculated at a rate of 0.15% per month on the minimum monthly balance. The interest payment for the month of October will be

A. $0.19   B. $0.57   C. $0.71   D. $0.92   E. $1.28

[VCAA 2006 FM]

Question 211

Grandpa invested in an ordinary perpetuity from which he receives a monthly pension of $584. The interest rate for the investment is 6.2% per annum. The amount Grandpa has invested in the perpetuity is closest to

A. $3600   B. $9420   C. $94 200   D. $43 400   E. $113 000

[VCAA 2006 FM]

Question 212

An item was purchased for a price of $825. The price included 10% GST (Goods and Services Tax). The amount of GST included in the price is

A. $8.25   B. $75.00   C. $82.50   D. $90.75   E. $125.00

[VCAA 2006 FM]

Question 213

A photocopier is depreciated by $0.04 for each copy it makes. Three years ago the photocopier was purchased for $48 000. Its depreciated value now is $21 000. The total number of copies made by the photocopier in the three years is

A. 108 000   B. 192 000   C. 276 000   D. 525 000   E. 675 000

[VCAA 2006 FM]

B4. Business-related mathematics

Question 214

A $2000 lounge suite was sold under a hire-purchase agreement. A deposit of $200 was paid. The balance was to be paid in 36 equal monthly instalments of $68.
The annual flat rate of interest applied to this agreement is

A. 10.0% B. 11.4% C. 12.0% D. 22.4% E. 36.0%

[VCAA 2006 FM]

Question 215

Mervyn bought a new lawn mower at a sale. First, there was a 20% discount from the original price. Then, an $80 trade-in for his old mower was subtracted from this reduced price. This left Mervyn with $368 to pay for the new lawn mower. The original price of the new lawn mower was

A. $468.00 B. $537.50 C. $540.00 D. $560.00 E. $580.00

[VCAA 2006 FM]

Question 216

The points on the graph to the right show the balance of an investment at the start of each quarter for a period of six years. The same rate of interest applied for these six years.

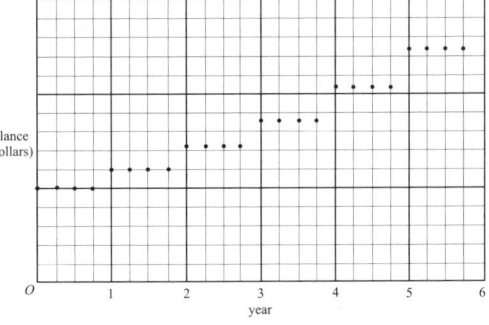

In relation to this investment, which one of the following statements is **true**?

A. interest is compounding annually and is credited annually
B. interest is compounding annually and is credited quarterly
C. interest is compounding quarterly and is credited quarterly
D. simple interest is paid on the opening balance and is credited annually
E. simple interest is paid on the opening balance and is credited quarterly

[VCAA 2006 FM]

Question 217

Jenny borrowed $18 000. She will fully repay the loan in five years with equal monthly payments. Interest is charged at the rate of 9.2% per annum, calculated monthly, on the reducing balance. The amount Jenny has paid off the principal immediately following the **tenth** repayment is

A. $1876.77 B. $2457.60 C. $3276.00 D. $3600.44 E. $3754.00

[VCAA 2006 FM]

B4. Business-related mathematics

Question 218

An agent charged $20 commission for selling a rare book for $500. What percentage of the selling price is the commission?

A. 4% B. 5% C. 20% D. 25% E. 40%

Question 219

A car is valued at $30 000 when new. Its value is depreciated by 25 cents for each kilometre it travels. The number of kilometres the car travels before its value depreciates to $8000 is

A. 32000 B. 55 000 C. 88 000 D. 120 000 E. 550 000

Question 220

A sum of money is invested in an account paying simple interest at a rate of 8% per annum. The total interest earned on this investment over 6 years is $27 000.
The sum of money invested is

A. $12 960 B. $45 000 C. $56 250
D. $202 500 E. $337 500

Question 221

The price of a property purchased in 2006 was $200 000.
Stamp duty was paid on this purchase according to the schedule below.

Price range	Rate
$0 – $20 000	1.4 per cent of the price of the property
$20 001 – $115 000	$280 plus 2.4 per cent of the price in excess of $20 000
$115 001 – $868 000	$2560 plus 6 per cent of the price in excess of $115 000
More than $868 000	5.5 per cent of the price

The amount of stamp duty paid was

A. $2560 B. $2800 C. $5100
D. $7660 E. $9460

[VCAA 2007 FM]

B4. Business-related mathematics

Question 222

A new kitchen in a restaurant cost $50 000. Its value is depreciated over time using the reducing balance method. The value of the kitchen in dollars at the end of each year for ten years is shown in the graph here.

Which one of the following statements is **true**?

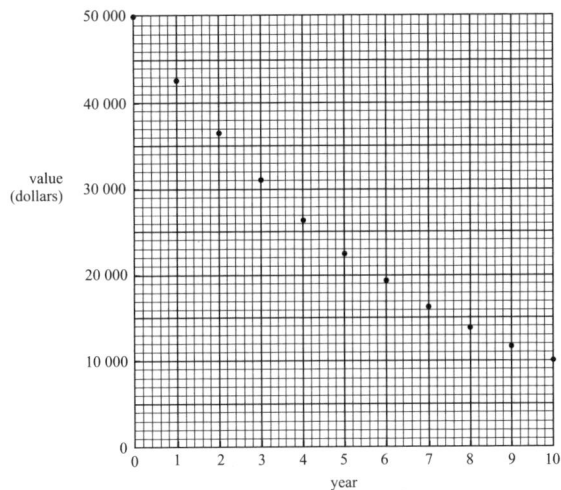

A. The kitchen depreciates by $4000 annually.
B. At the end of five years, the kitchen's value is less than $20 000.
C. The reducing balance depreciation rate is less than 5% per annum.
D. The annual depreciation rate increases over time.
E. The amount of depreciation each year decreases over time.

[VCAA 2007 FM]

Question 223

$10 000 is invested at a rate of 10% per annum compounding half yearly. The value, in dollars, of this investment after five years, is given by

A. $10\,000 \times 0.10 \times 5$
B. $10\,000 \times 0.05 \times 10$
C. $10\,000 \times 0.05^{10}$
D. $10\,000 \times 1.05^{10}$
E. $10\,000 \times 1.10^{5}$

[VCAA 2007 FM]

Question 224

At the start of each year Joe's salary increases to take inflation into account. Inflation averaged 2% per annum last year and 3% per annum the year before that. Joe's salary this year is $42 000.
Joe's salary two years ago, correct to the nearest dollar, would have been

A. $39 900
B. $39 925
C. $39 926
D. $39 976
E. $39 977

[VCAA 2007 FM]

65

B4. Business-related mathematics

Question 225

Brad investigated the cost of buying a $720 washing machine under a hire purchase agreement. A deposit of $180 is required and the balance will be paid in 24 equal monthly repayments. A flat interest rate of 12% per annum applies to the balance.
Brad correctly calculated the monthly repayment to be

A. $22.50 B. $25.20 C. $26.10
D. $27.90 E. $29.70

[VCAA 2007 FM]

Question 226

Petra borrowed $250 000 to buy a home. The interest rate is 7% per annum, calculated monthly on the reducing balance over the life of the loan. She will fully repay the loan over 20 years with equal monthly instalments.
The total amount of interest she will pay on the loan is closest to

A. $215 000 B. $266 000 C. $281 000
D. $350 000 E. $465 000

[VCAA 2007 FM]

B5. Networks and decision mathematics

Question 227

The sum of the degrees of all the vertices in this network is

A. 6 **B.** 7 **C.** 8
D. 9 **E.** 10

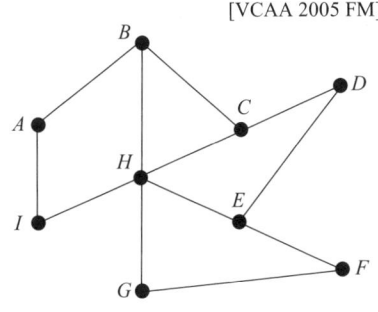

[VCAA 2005 FM]

Question 228

For the network shown, an Euler **path** can be found
A. without altering the network.
B. by adding an edge that joins A to H.
C. by adding an edge that joins C to F.
D. by removing the edge that joins B to C.
E. by removing the edge that joins D to E.

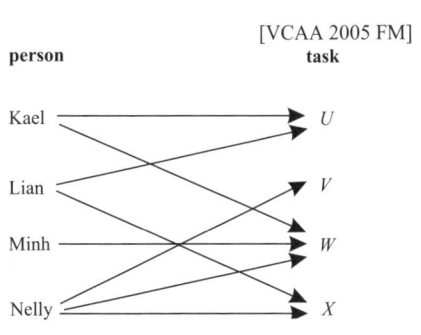

[VCAA 2005 FM]

Question 229

The bipartite graph here represents the tasks that four people are able to undertake.

The matrix representation for this task allocation is

A.
$$\begin{array}{c} \\ \text{Kael} \\ \text{Lian} \\ \text{Minh} \\ \text{Nelly} \end{array} \begin{array}{cccc} U & V & W & X \\ \end{array} \begin{bmatrix} 1 & 1 & 0 & 0 \\ 1 & 0 & 0 & 1 \\ 0 & 0 & 1 & 0 \\ 0 & 1 & 1 & 1 \end{bmatrix}$$

B.
$$\begin{array}{c} \\ \text{Kael} \\ \text{Lian} \\ \text{Minh} \\ \text{Nelly} \end{array} \begin{array}{cccc} U & V & W & X \\ \end{array} \begin{bmatrix} 1 & 0 & 1 & 0 \\ 1 & 0 & 0 & 1 \\ 0 & 0 & 1 & 0 \\ 0 & 1 & 1 & 1 \end{bmatrix}$$

C.
$$\begin{array}{c} \\ \text{Kael} \\ \text{Lian} \\ \text{Minh} \\ \text{Nelly} \end{array} \begin{array}{cccc} U & V & W & X \\ \end{array} \begin{bmatrix} 1 & 0 & 1 & 0 \\ 1 & 0 & 0 & 0 \\ 0 & 0 & 1 & 0 \\ 0 & 1 & 1 & 1 \end{bmatrix}$$

D.
$$\begin{array}{c} \\ \text{Kael} \\ \text{Lian} \\ \text{Minh} \\ \text{Nelly} \end{array} \begin{array}{cccc} U & V & W & X \\ \end{array} \begin{bmatrix} 1 & 0 & 1 & 0 \\ 1 & 0 & 0 & 1 \\ 0 & 0 & 1 & 1 \\ 0 & 1 & 1 & 1 \end{bmatrix}$$

E.
$$\begin{array}{c} \\ \text{Kael} \\ \text{Lian} \\ \text{Minh} \\ \text{Nelly} \end{array} \begin{array}{cccc} U & V & W & X \\ \end{array} \begin{bmatrix} 1 & 0 & 1 & 0 \\ 1 & 1 & 0 & 1 \\ 0 & 0 & 1 & 0 \\ 0 & 1 & 1 & 1 \end{bmatrix}$$

[VCAA 2005 FM]

B5. Networks and decision mathematics

Question 230

The network diagram here shows distances (in metres) between different points in a garden, A, B, C, D, E, F, G, H and I. Also shown are the distances from a tank (T) and these points. Pipes will carry water from the tank to each of these nine points. In the network diagrams below, the water pipes are indicated in bold. Which of these network diagrams shows the minimum length of water pipe required to connect the tank (T) to each of the nine points?

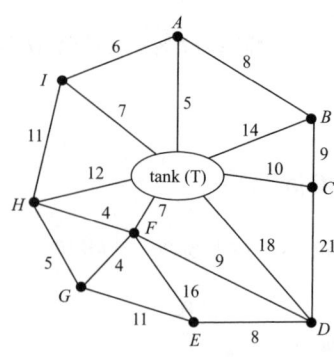

A.

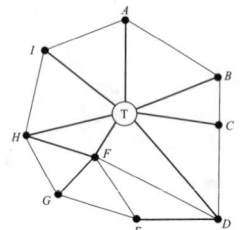

B.

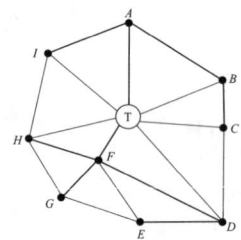

C.

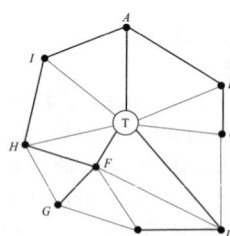

D.

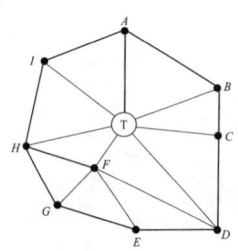

E.

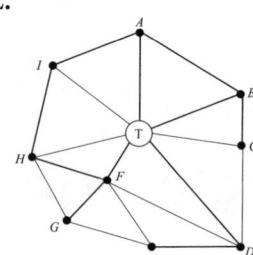

[VCAA 2005 FM]

Question 231

A connected planar graph has 10 vertices and 15 edges. A number of edges are removed to leave a connected graph with 10 vertices and 3 faces. The number of edges that were removed is

A. 4   B. 5   C. 6   D. 7   E. 8

[VCAA 2005 FM]

Question 232

On the directed graph shown on the next page, the values on the edges give the maximum flow between nodes in the direction of the arrows. Five cuts have been made on the diagram. Which cut allows you to find the maximum flow from point X to point Y?

B5. Networks and decision mathematics

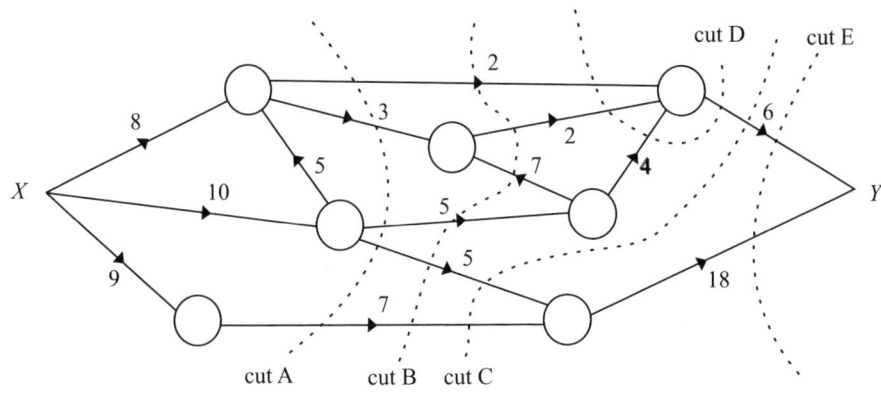

A. cut A B. cut B C. cut C D. cut D E. cut E

[VCAA 2005 FM]

Question 233

The diagram represents a network of road connections between the four towns, *J*, *K*, *L* and *M*. The map that corresponds to the roads that are represented by the network diagram could be

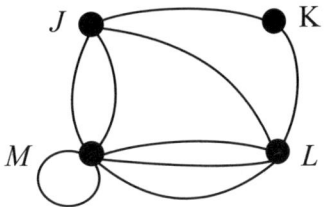

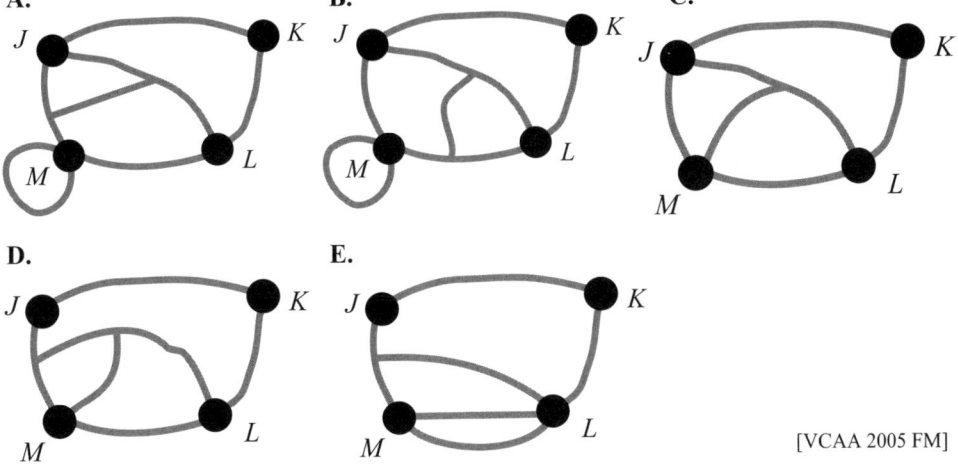

[VCAA 2005 FM]

B5. Networks and decision mathematics

Question 234

The activities and their completion times (in hours) that are needed to complete a project are shown in the network below. For this project, the minimum time to complete the whole project would be increased if one of the activities was delayed.

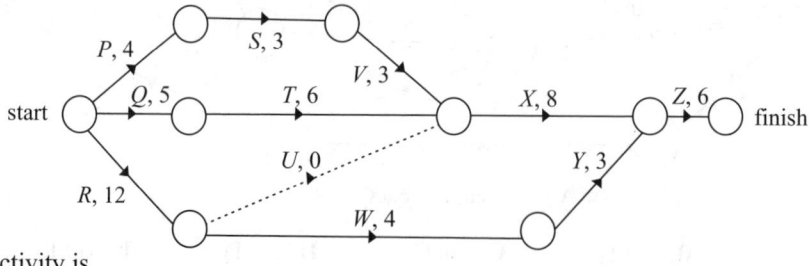

This activity is
A. P   B. Q   C. R   D. T   E. W

[VCAA 2005 FM]

Question 235

The matrix shows the roads between five towns, Q, R, S, T, U and the distances in kilometres between the towns. A dash (–) indicates that there is no direct connection between two particular towns. The shortest route, in kilometres, between Q and T is

A. 10   B. 21   C. 23   D. 25   E. 27

[VCAA 2005 FM]

Question 236

The number of vertices with an odd degree in the network shown is

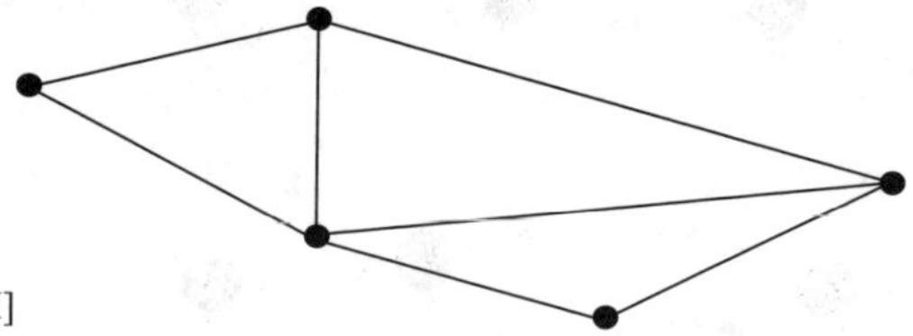

A. 1   B. 2
C. 3   D. 4
E. 5

[VCAA 2006 FM]

Question 237

The following directed graph represents a series of one-way streets with intersections numbered as nodes 1 - 8. All intersections can be reached from

A. intersection 4
B. intersection 5
C. intersection 6
D. intersection 7
E. intersection 8

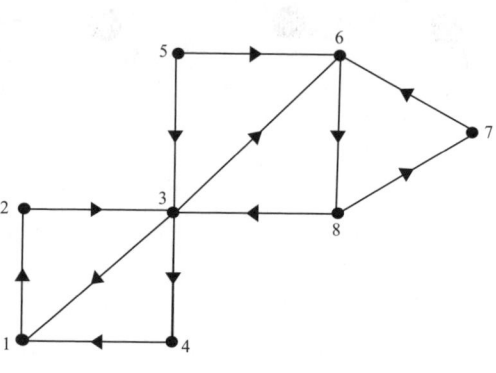

[VCAA 2006 FM]

70

B5. Networks and decision mathematics

Question 238

Which one of the following statements is **true** regarding the network shown?

A. *ABCDEFG* is a Hamiltonian circuit.
B. Only one Hamiltonian path exists.
C. *CBAGFEDC* is an Eulerian circuit.
D. At least two Eulerian paths exist.
E. There are no circuits.

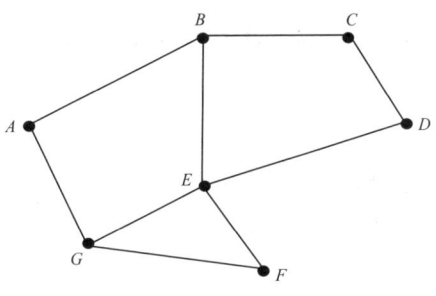

[VCAA 2006 FM]

Question 239

The minimal spanning tree for the network shown will include the edge that has a weight of

A. 3
B. 6
C. 8
D. 9
E. 10

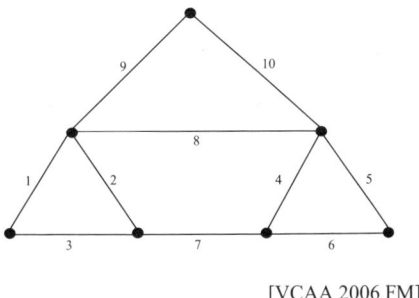

[VCAA 2006 FM]

Question 240

For a particular project there are ten activities that must be completed. These activities and their immediate predecessors are given in the table below.

Activity	Immediate predecessors
A	–
B	–
C	–
D	A
E	B
F	D, E
G	C
H	C
I	F, G
J	H, I

B5. Networks and decision mathematics

A directed graph that could represent this project is

A.   **B.**

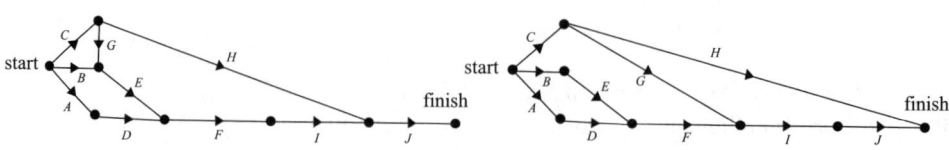

C.   **D.**

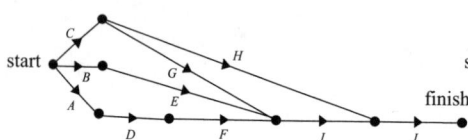   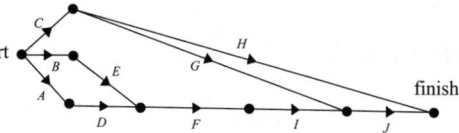

E.

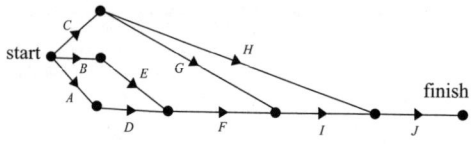

[VCAA 2006 FM]

Question 241

In the directed graph shown here, the weight of each edge is non-zero. The capacity of the cut shown is

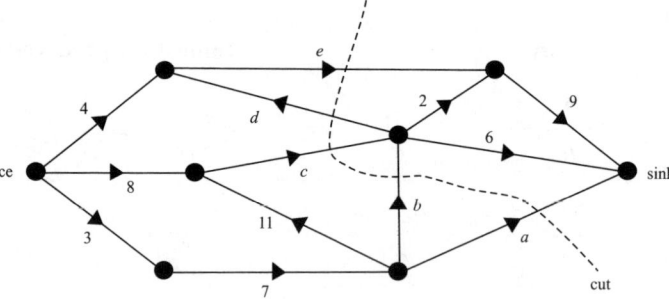

A. $a + b + c + d + e$
B. $a + c + d + e$
C. $a + b + c + e$
D. $a + b + c - d + e$
E. $a - b + c - d + e$

[VCAA 2006 FM]

Question 242

A **complete** graph with six vertices is drawn. This network would best represent

A. the journey of a paper boy who delivers to six homes covering the minimum distance.
B. the cables required to connect six houses to pay television that minimises the length of cables needed.
C. a six-team basketball competition where all teams play each other once.

B5. Networks and decision mathematics

D. a project where six tasks must be performed between the start and finish.
E. the allocation of different assignments to a group of six students.

[VCAA 2006 FM]

Question 243

Euler's formula, relating vertices, faces and edges, does **not** apply to which one of the following graphs?

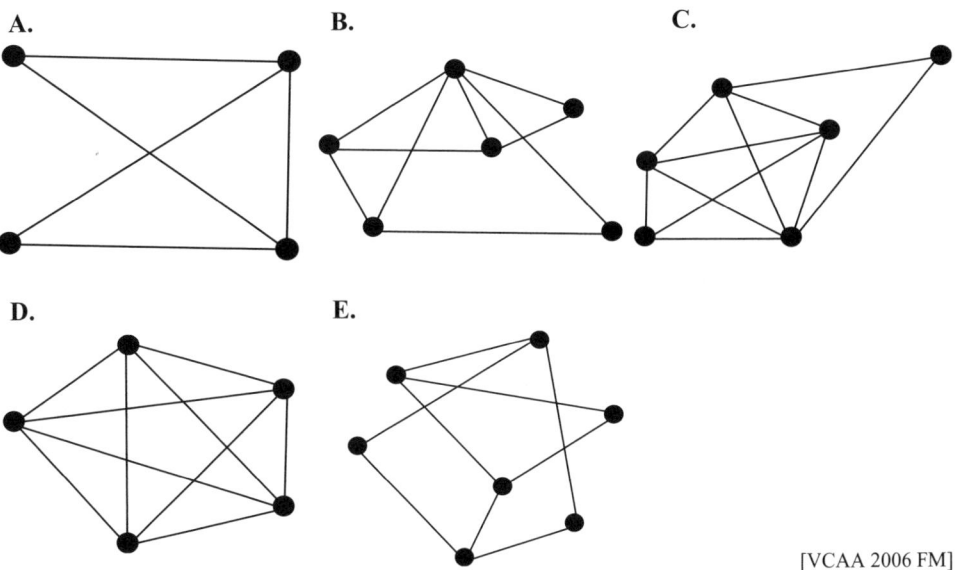

[VCAA 2006 FM]

Question 244

The network here shows the activities and their completion times (in hours) that are needed to complete a project.

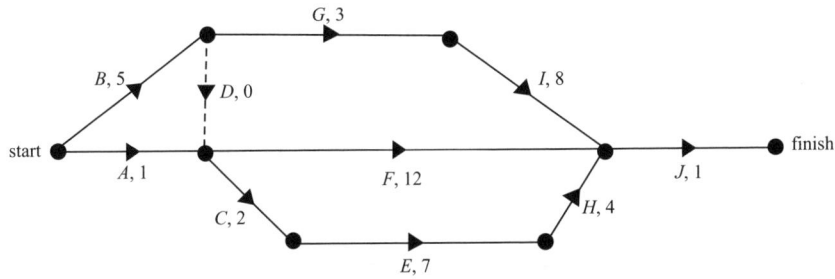

The project is to be crashed by reducing the completion time of **one** activity only. This will reduce the completion time of the project by a maximum of

A. 1 hour **B.** 2 hours **C.** 3 hours **D.** 4 hours **E.** 5 hours

[VCAA 2006 FM]

B5. Networks and decision mathematics

Question 245

A mathematical term that could **not** be used to describe the graph shown here is

A. complete.
B. planar.
C. simple.
D. undirected.
E. tree.

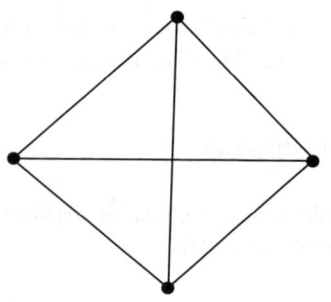

[VCAA 2007 FM]

Question 246

A connected planar graph has 12 edges. This graph could have

A. 5 vertices and 6 faces.
B. 5 vertices and 8 faces.
C. 6 vertices and 8 faces.
D. 6 vertices and 9 faces.
E. 7 verticcs and 9 faces.

[VCAA 2007 FM]

Question 247

Consider the following graph.
An adjacency matrix that could be used to represent this graph is

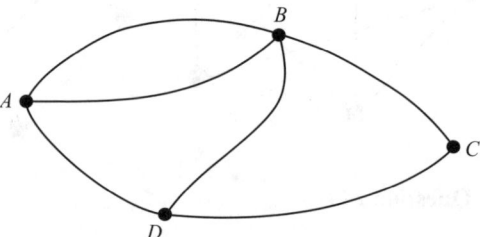

[VCAA 2007 FM]

A.
$$\begin{bmatrix} 0 & 2 & 0 & 1 \\ 2 & 0 & 1 & 1 \\ 0 & 1 & 0 & 1 \\ 1 & 1 & 1 & 0 \end{bmatrix}$$

B.
$$\begin{bmatrix} 0 & 2 & 0 & 1 \\ 0 & 0 & 1 & 1 \\ 0 & 0 & 0 & 1 \\ 0 & 0 & 0 & 0 \end{bmatrix}$$

C.
$$\begin{bmatrix} 0 & 1 & 0 & 1 \\ 2 & 0 & 0 & 1 \\ 0 & 1 & 0 & 1 \\ 1 & 1 & 1 & 0 \end{bmatrix}$$

D.
$$\begin{bmatrix} 0 & 2 & 0 & 1 \\ 0 & 1 & 1 & 1 \\ 0 & 1 & 1 & 1 \\ 0 & 1 & 1 & 1 \end{bmatrix}$$

E.
$$\begin{bmatrix} 1 & 2 & 0 & 1 \\ 2 & 1 & 0 & 1 \\ 0 & 1 & 1 & 0 \\ 0 & 0 & 1 & 1 \end{bmatrix}$$

[VCAA 2007 FM]

B5. Networks and decision mathematics

Question 248

The following network shows the distances, in kilometres, along a series of roads that connect town A to town B.

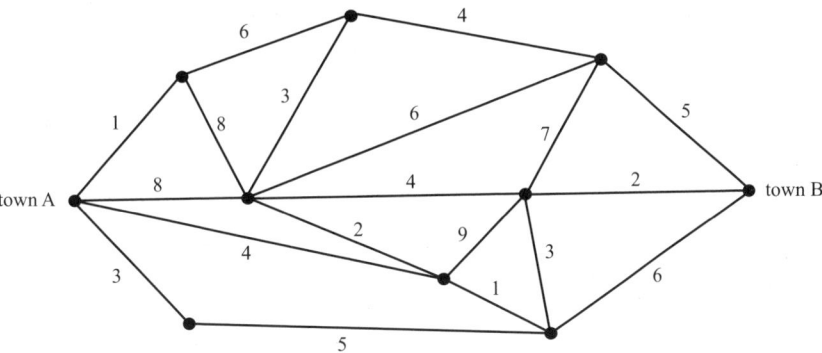

The shortest distance, in kilometres, to travel from town A to town B is

A. 9   B. 10   C. 11   D. 12   E. 13

[VCAA 2007 FM]

Questions 249 and 250 relate to the following information.

The following network shows the activities that are needed to complete a project and their completion times (in hours)

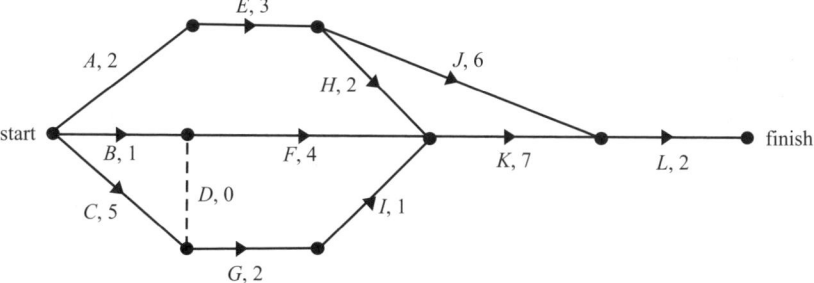

Question 249

Which one of the following statements regarding this project is false?

A. Activities A, B and C all have the same earliest start time.
B. There is only one critical path for this project.
C. Activity J may start later than activity H.
D. The shortest path gives the minimum time for project completion.
E. Activity L must be on the critical path.

[VCAA 2007 FM]

B5. Networks and decision mathematics

Question 250

The earliest start time for activity L, in hours, is

A. 11 **B.** 12 **C.** 14 **D.** 15 **E.** 16

[VCAA 2007 FM]

Question 251

The minimal spanning tree for the network shown here includes two edges with weightings x and y.
The length of the minimal spanning tree is 19. The values of x and y could be

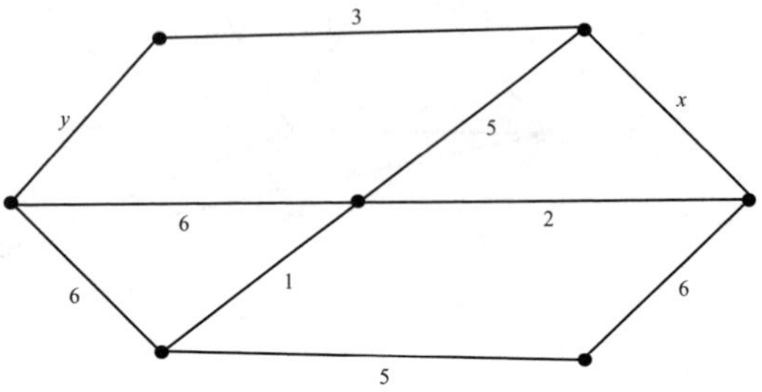

A. $x = 1$ and $y = 7$
B. $x = 2$ and $y = 5$
C. $x = 3$ and $y = 5$
D. $x = 4$ and $y = 5$
E. $x = 5$ and $y = 6$

[VCAA 2007 FM]

Questions 252 and 253 relate to the following information.

There are five teams, A, B, C, D and E, in a volleyball competition. Each team played each other team once in 2007. The results are summarised in the directed graph shown here. An arrow from A to E signifies that A defeated E.

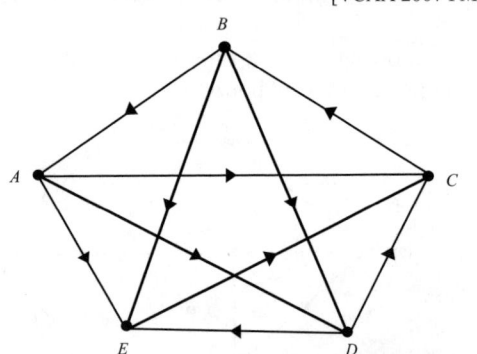

Question 252

In 2007, the team that had the highest number of two-step dominances was

A. team A **B.** team B **C.** team C **D.** team D **E.** team E

[VCAA 2007 FM]

Question 253

In 2008, two new teams, F and G, will join the competition. As in 2007, each team will play every other team once. Compared to 2007, the number of extra games that will be played in 2008 will be

A. 10 **B.** 11 **C.** 12 **D.** 21 **E.** 42

[VCAA 2007 FM]

B6. Matrices

Question 254

The matrix $\begin{bmatrix} 12 & 36 \\ 0 & 24 \end{bmatrix}$ is equal to

A. $12\begin{bmatrix} 0 & 3 \\ 0 & 2 \end{bmatrix}$ B. $12\begin{bmatrix} 1 & 3 \\ 0 & 2 \end{bmatrix}$    C. $12\begin{bmatrix} 0 & 24 \\ -12 & 12 \end{bmatrix}$

D. $12\begin{bmatrix} 0 & 24 \\ 0 & 12 \end{bmatrix}$ E. $12\begin{bmatrix} 1 & 3 \\ -12 & 2 \end{bmatrix}$

[VCAA 2006 FM]

Question 255

Let $A = \begin{bmatrix} -2 \\ 0 \end{bmatrix}$, $B = [0 \ 9]$ and $C = [2]$. Using these matrices, the matrix product that is **not** defined is

A. AB B. AC C. BA D. BC E. CB

[VCAA 2006 FM]

Question 256

Let $A = \begin{bmatrix} 1 & 0 \\ 0 & 1 \end{bmatrix}$, $B = \begin{bmatrix} 2 & 1 \\ 1 & 0 \end{bmatrix}$ and $C = \begin{bmatrix} 1 & -1 \\ -1 & 1 \end{bmatrix}$. Then $A^3(B-C)$ equals

A. $\begin{bmatrix} 1 & 2 \\ 2 & -1 \end{bmatrix}$    B. $\begin{bmatrix} 1 & 0 \\ 0 & -1 \end{bmatrix}$ C. $\begin{bmatrix} 3 & 6 \\ 6 & -3 \end{bmatrix}$

D. $\begin{bmatrix} 3 & 0 \\ 0 & -3 \end{bmatrix}$ E. $\begin{bmatrix} 5 & 10 \\ 10 & -5 \end{bmatrix}$

[VCAA 2006 FM]

B6. Matrices

Question 257

Three teams, Blue (*B*), Green (*U*) and Red (*R*), compete for three different sporting competitions. The table shows the competition winners for the past three years.

	Athletics	Cross country	Swimming
2004	Green	Green	Blue
2005	Green	Red	Blue
2006	Blue	Green	Blue

A matrix that shows the **total number** of competitions won by **each** of the three teams in each of these three years could be

A. $\begin{array}{c} \\ 2004 \\ 2005 \\ 2006 \end{array} \begin{array}{ccc} B & G & R \\ \left[\begin{matrix} 1 & 1 & 1 \\ 1 & 1 & 1 \\ 1 & 1 & 1 \end{matrix}\right] \end{array}$

B. $\begin{array}{c} \\ 2004 \\ 2005 \\ 2006 \end{array} \begin{array}{ccc} B & G & R \\ \left[\begin{matrix} 1 & 2 & 0 \\ 1 & 1 & 1 \\ 2 & 1 & 0 \end{matrix}\right] \end{array}$

C. $\begin{array}{c} \\ 2004 \\ 2005 \\ 2006 \end{array} \begin{array}{ccc} B & G & R \\ \left[\begin{matrix} 1 & 2 & 0 \\ 1 & 1 & 1 \\ 1 & 1 & 0 \end{matrix}\right] \end{array}$

D. $\begin{array}{c} \\ 2004 \\ 2005 \\ 2006 \end{array} \begin{array}{ccc} B & G & R \\ \left[\begin{matrix} 4 & 0 & 0 \\ 0 & 4 & 0 \\ 0 & 0 & 4 \end{matrix}\right] \end{array}$

E. $\begin{array}{c} \\ 2004 \\ 2005 \\ 2006 \end{array} \begin{array}{ccc} B & G & R \\ \left[\begin{matrix} 2 & 1 & 0 \\ 2 & 0 & 1 \\ 0 & 3 & 0 \end{matrix}\right] \end{array}$

[VCAA 2006 FM]

Question 258

A company makes Regular (*R*), Queen (*Q*) and King (*K*) size beds. Each bed comes in either the Classic style or the more expensive Deluxe style. The price of each style of bed, in dollars, is listed in a price matrix *P*, where

$$P = \begin{array}{c} \\ \end{array} \begin{array}{ccc} R & Q & K \\ \left[\begin{matrix} 145 & 210 & 350 \\ 185 & 270 & 410 \end{matrix}\right] \end{array} \begin{array}{c} \text{Classic} \\ \text{Deluxe} \end{array}$$

The company wants to increase the price of all beds. A new price matrix, listing the increased prices of the beds, can be generated from *P* by forming a matrix product with the matrix, *M*, where $M = \begin{bmatrix} 1.2 & 0 \\ 0 & 1.35 \end{bmatrix}$. This new price matrix is

A. $\begin{bmatrix} 145 & 210 & 350 \\ 185 & 270 & 410 \end{bmatrix}$

B. $\begin{bmatrix} 234.90 & 340.20 & 567 \\ 299.70 & 437.40 & 664.20 \end{bmatrix}$

B6. Matrices

C. $\begin{bmatrix} 174 & 252 & 420 \\ 222 & 324 & 492 \end{bmatrix}$ D. $\begin{bmatrix} 174 & 252 & 420 \\ 249.75 & 364.50 & 553.50 \end{bmatrix}$

E. $\begin{bmatrix} 195.75 & 283.50 & 472.50 \\ 249.75 & 364.50 & 553.50 \end{bmatrix}$

[VCAA 2006 FM]

Question 259

If $A = \begin{bmatrix} 1 & 3 \\ 6 & 4 \\ 0 & 0 \end{bmatrix}$ and the matrix product $XA = \begin{bmatrix} 4 & 1 \\ 1 & 4 \\ 3 & 5 \end{bmatrix}$, then the order of matrix X is

A. (2×2) B. (2×3) C. (3×1) D. (3×2) E. (3×3)

[VCAA 2006 FM]

Question 260

How many of the following five sets of simultaneous linear equations have a unique solution?

$4x + 2y = 10$	$x = 0$	$x - y = 3$	$2x + y = 5$	$x = 8$
$2x + y = 5$	$x + y = 6$	$x + y = 3$	$2x + y = 10$	$y = 2$

A. 1   B. 2   C. 3   D. 4   E. 5

[VCAA 2006 FM]

Question 261

Australians go on holidays either within Australia or overseas. Market research shows that
- 95% of those who had their last holiday in Australia said that their next holiday would be in Australia
- 20% of those who had their last holiday overseas said that their next holiday would also be overseas.

A transition matrix that could be used to describe this situation is

A. $\begin{bmatrix} 0.95 \\ 0.20 \end{bmatrix}$ B. $\begin{bmatrix} 0.95 \\ 0.05 \end{bmatrix} + \begin{bmatrix} 0.20 \\ 0.80 \end{bmatrix}$ C. $\begin{bmatrix} 0.95 & 0.95 \\ 0.20 & 0.20 \end{bmatrix}$

D. $\begin{bmatrix} 0.95 & 0.20 \\ 0.05 & 0.80 \end{bmatrix}$ E. $\begin{bmatrix} 0.95 & 0.80 \\ 0.05 & 0.20 \end{bmatrix}$

[VCAA 2006 FM]

B6. Matrices

Question 262

A large population of birds lives on a remote island. Every night each bird settles at either location A or location B. It was found on the first night that the number of birds at each location was the same. On each subsequent night, a percentage of birds changed the location at which they settled. This movement of birds is described by the transition matrix

$$\begin{array}{c} & \begin{array}{cc} A & B \end{array} \\ \begin{array}{c} A \\ B \end{array} & \begin{bmatrix} 0.8 & 0 \\ 0.2 & 1 \end{bmatrix} \end{array}$$

Assume this pattern of movement continues. In the long term, the number of birds that settle at location A will

A. not change.
B. gradually decrease to zero.
C. eventually settle at around 20% of the island's bird population.
D. eventually settle at around 80% of the island's bird population.
E. gradually increase.

[VCAA 2006 FM]

Question 263

The matrix sum $\begin{bmatrix} 0 & -4 \\ 2 & 5 \end{bmatrix} + \begin{bmatrix} 5 & 4 \\ -2 & 2 \end{bmatrix}$ is equal to

A. $\begin{bmatrix} 5 & 0 \\ 0 & 7 \end{bmatrix}$
B. $\begin{bmatrix} 0 & 0 \\ 0 & 7 \end{bmatrix}$
C. $\begin{bmatrix} 5 & -4 \\ 0 & 7 \end{bmatrix}$

D. $\begin{bmatrix} 0 & 5 & -4 & 4 \\ 2 & -2 & 5 & 2 \end{bmatrix}$
E. $\begin{bmatrix} 0 & -4 & 5 & 4 \\ 2 & 5 & -2 & 2 \end{bmatrix}$

[VCAA 2007 FM]

Question 264

The number of tourists visiting three towns, Oldtown, Newtown and Twixtown, was recorded for three years. The data is summarised in the table below.

	2004	2005	2006
Old town	975	1002	1390
Newtown	2105	1081	1228
Twixtown	610	1095	1380

B6. Matrices

The 3 x 1 matrix that could be used to show the number of tourists visiting the three towns in the year 2005 is

A. [975 1002 1390] **B.** [1002 1081 1095] **C.** $\begin{bmatrix} 975 \\ 1002 \\ 1390 \end{bmatrix}$

D. $\begin{bmatrix} 1002 \\ 1081 \\ 1095 \end{bmatrix}$ **E.** $\begin{bmatrix} 975 & 1002 & 1390 \\ 2105 & 1081 & 1228 \\ 610 & 1095 & 1380 \end{bmatrix}$

[VCAA 2007 FM]

Question 265

If $A = \begin{bmatrix} 8 & 4 \\ 5 & 3 \end{bmatrix}$ and the product $AX = \begin{bmatrix} 5 & 6 \\ 8 & 10 \end{bmatrix}$ then X is

A. $\begin{bmatrix} 24 & -14 \\ 13 & -7.5 \end{bmatrix}$ **B.** $\begin{bmatrix} -4.25 & -5.5 \\ 9.75 & 12.5 \end{bmatrix}$ **C.** $\begin{bmatrix} -3.75 & 7 \\ -6.5 & 12 \end{bmatrix}$

D. $\begin{bmatrix} 25 & 11 \\ -19.5 & -8.5 \end{bmatrix}$ **E.** $\begin{bmatrix} 0.625 & 1.5 \\ 1.6 & 3.333 \end{bmatrix}$

[VCAA 2007 FM]

Question 266

Consider the following system of three simultaneous linear equations.
$2x + z = 5$
$x - 2y = 0$
$y - z = -1$

This system of equations can be written in matrix form as

A. $\begin{bmatrix} 2 & 1 \\ 1 & -2 \\ 1 & -1 \end{bmatrix} \begin{bmatrix} x \\ y \\ z \end{bmatrix} = \begin{bmatrix} 5 \\ 0 \\ -1 \end{bmatrix}$ **B.** $\begin{bmatrix} 2 & 0 & 1 \\ 1 & -2 & 0 \\ 0 & 1 & -1 \end{bmatrix} \begin{bmatrix} x \\ y \\ z \end{bmatrix} = \begin{bmatrix} 5 \\ 0 \\ -1 \end{bmatrix}$ **C.** $\begin{bmatrix} 2 & 1 & 5 \\ 1 & -2 & 0 \\ 1 & -1 & -1 \end{bmatrix} \begin{bmatrix} x \\ y \\ z \end{bmatrix} = \begin{bmatrix} 5 \\ 0 \\ -1 \end{bmatrix}$

D. $\begin{bmatrix} 2 & 1 & 0 \\ 1 & -2 & 0 \\ 1 & -1 & 0 \end{bmatrix} \begin{bmatrix} x \\ y \\ z \end{bmatrix} = \begin{bmatrix} 5 \\ 0 \\ -1 \end{bmatrix}$ **E.** $\begin{bmatrix} 2 & 1 & 5 \\ 1 & -2 & 0 \\ 1 & -1 & -1 \end{bmatrix} \begin{bmatrix} 5 \\ 0 \\ -1 \end{bmatrix} = \begin{bmatrix} x \\ y \\ z \end{bmatrix}$

[VCAA 2007 FM]

B6. Matrices

Question 267

An international mathematics competition is conducted in three sections - Junior, Intermediate and Senior. There are money prizes for gold, silver and bronze levels of achievement in each of these sections. Table 1 shows the number of students who were awarded prizes in each section.

Number of students awarded prizes

Achievement level	Junior	Intermediate	Senior
Gold	21	12	10
Silver	16	18	14
Bronze	21	26	24

The value, in dollars, of each prize.

Achievement level	Junior	Intermediate	Senior
Gold	75	100	200
Silver	25	40	80
Bronze	10	15	20

A matrix product that gives the total value of all the Silver prizes that were awarded is

A. $[25 \ 40 \ 80] \begin{bmatrix} 16 \\ 18 \\ 14 \end{bmatrix}$
B. $[25 \ 40 \ 80][16 \ 18 \ 14]$

C. $\begin{bmatrix} 16 \\ 18 \\ 14 \end{bmatrix} \begin{bmatrix} 25 \\ 40 \\ 80 \end{bmatrix}$
D. $\begin{bmatrix} 16 \\ 18 \\ 14 \end{bmatrix} [25 \ 40 \ 80]$
E. $[100 \ 40 \ 15] \begin{bmatrix} 16 \\ 18 \\ 14 \end{bmatrix}$

[VCAA 2007 FM]

Question 268

A colony of fruit bats feeds nightly at three different locations, A, B and C.
Initially. the number of bats from the colony feeding at each of the locations was as follows.

Location	Number of bats
A	1568
B	1105
C	894

B6. Matrices

The bats change feeding locations according to the following transition matrix T.

$$T = \begin{array}{c} \text{this night} \\ \begin{array}{ccc} A & B & C \end{array} \\ \begin{bmatrix} 0.8 & 0.1 & 0.2 \\ 0.1 & 0.6 & 0.1 \\ 0.1 & 0.3 & 0.7 \end{bmatrix} \begin{array}{c} A \\ B \\ C \end{array} \text{ next night} \end{array}$$

If this pattern of feeding continues, the number of bats feeding at location A in the long term will be closest to

A. 1254   B. 1543   C. 1568   D. 1605   E. 1725

[VCAA 2007 FM]

Question 269

Each year, a family always goes on its holiday to one of three places; Portland (P), Quambatook (Q) or Rochester (R).
They never go to the same place two years in a row. For example, if they went to Portland one year, they would not go to Portland the next year; they would go to Quambatook or Rochester instead.
A transition matrix that can be used to model this situation is

A.
$$\text{Next year} \begin{array}{c} \\ P \\ Q \\ R \end{array} \begin{array}{c} \text{This year} \\ \begin{array}{ccc} P & Q & R \end{array} \\ \begin{bmatrix} 1 & 0 & 0 \\ 0 & 1 & 0 \\ 0 & 0 & 1 \end{bmatrix} \end{array}$$

B.
$$\text{Next year} \begin{array}{c} \\ P \\ Q \\ R \end{array} \begin{array}{c} \text{This year} \\ \begin{array}{ccc} P & Q & R \end{array} \\ \begin{bmatrix} 0 & 1 & 0 \\ 0.9 & 0 & 0 \\ 0.1 & 0 & 1 \end{bmatrix} \end{array}$$

C.
$$\text{Next year} \begin{array}{c} \\ P \\ Q \\ R \end{array} \begin{array}{c} \text{This year} \\ \begin{array}{ccc} P & Q & R \end{array} \\ \begin{bmatrix} 0 & 0.3 & 0.8 \\ 0.3 & 0.4 & 0.2 \\ 0.7 & 0.3 & 0 \end{bmatrix} \end{array}$$

D.
$$\text{Next year} \begin{array}{c} \\ P \\ Q \\ R \end{array} \begin{array}{c} \text{This year} \\ \begin{array}{ccc} P & Q & R \end{array} \\ \begin{bmatrix} 0 & 0.2 & 0 \\ 0.3 & 0 & 0.8 \\ 0.5 & 0.6 & 0 \end{bmatrix} \end{array}$$

E.
$$\text{Next year} \begin{array}{c} \\ P \\ Q \\ R \end{array} \begin{array}{c} \text{This year} \\ \begin{array}{ccc} P & Q & R \end{array} \\ \begin{bmatrix} 0 & 0 & 0.1 \\ 0.5 & 0 & 0.9 \\ 0.5 & 1 & 0 \end{bmatrix} \end{array}$$

[VCAA 2007 FM]

Question 270

Kerry sat for a multiple-choice test consisting of six questions. Each question had four alternative answers, A, B, C or D. He selected D for his answer to the first question. He then determined the answers to the remaining questions by following the transition matrix:

B6. Matrices

$$\text{Next question} \quad \begin{array}{c} \\ A \\ B \\ C \\ D \end{array} \overset{\begin{array}{cccc} A & B & C & D \end{array}}{\begin{bmatrix} 1 & 0 & 1 & 0 \\ 0 & 0 & 0 & 1 \\ 0 & 1 & 0 & 0 \\ 0 & 0 & 0 & 0 \end{bmatrix}}$$

The answers that he gave to the six test questions, starting with D, were

A.

Question	1	2	3	4	5	6
Answer	D	B	C	A	D	B

B.

Question	1	2	3	4	5	6
Answer	D	B	C	A	A	A

C.

Question	1	2	3	4	5	6
Answer	D	B	C	A	C	A

D.

Question	1	2	3	4	5	6
Answer	D	A	C	B	D	D

E.

Question	1	2	3	4	5	6
Answer	D	C	B	A	B	C

[VCAA 2007 FM]

Question 271

Matrix M is a 3×4 matrix. Matrix P has five rows. N is another matrix.
If the matrix product

$$M(NP) = \begin{bmatrix} 4 & 1 & 7 & 2 \\ 0 & 9 & 7 & 4 \\ 4 & 3 & 3 & 1 \end{bmatrix} \text{ then the order of matrix } N \text{ is}$$

A. 3×5 **B.** 5×3 **C.** 4×5 **D.** 5×4 **E.** 5×5

[VCAA 2007 FM]

C1. Core – extended responses

Question 272

This table gives the heights (m) and weights (kg) of a sample of nine people:

height (m)	weight (kg)
1.65	68
1.68	**63**
1.72	79
1.73	65
1.74	70
1.77	79
1.78	**81**
1.86	77
1.92	88

a. On the scatterplot shown, the points representing the data for seven of these people have been plotted with height on the horizontal axis and weight on the vertical axis. Plot points representing the data for the remaining two people (shown in bold in the table) on the scatterplot.

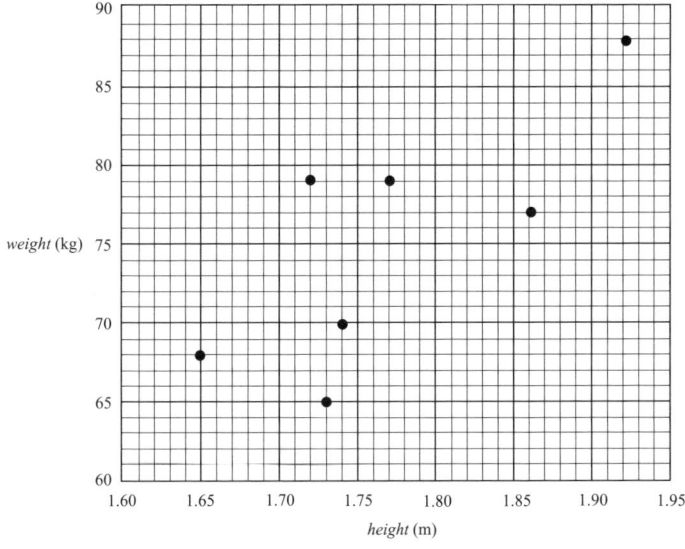

[1 mark]

b. Determine the equation of the least squares regression line that fits the data in the table. Use *height* as the independent variable and *weight* as the dependent variable. Complete the regression equation by writing the appropriate values in the boxes, correct to one decimal place.

weight = ☐ + ☐ × height

[2 marks]

C1. Core – extended responses

c. The coefficient of determination for this data is 0.61. Complete the following sentence by filling in the missing information.

For this sample, 61% of the variation in the ☐ of the people can be explained by the variation in their ☐.

[1 mark]
[VCAA 2004 FM]

Question 273

Body Mass Index (BMI) is defined as $BMI = \dfrac{weight}{(height)^2}$ where *weight* is measured in kilograms and *height* in metres.

a. Determine the *Body Mass Index* of a person who weighs 66 kg and who is 1.69 m tall. Write your answer correct to one decimal place.

[1 mark]

The *BMI* for each person in a sample of 17 males and 21 females is recorded in the next table.

Body Mass Index	
males	females
31.4	27.0
30.1	26.9
26.8	25.2
25.7	24.6
25.5	24.2
25.5	24.2
23.6	23.4
23.3	23.4
22.5	22.8
22.4	22.5
22.3	22.4
22.0	21.8
21.8	21.5
21.6	21.4
21.1	20.9
20.9	20.6
20.6	20.3
	20.1
	19.9
	18.8
	17.5

b. Write the range of the *BMI* data for **males** in the sample.

[1 mark]

C1. Core – extended responses

c. A *BMI* greater than 25 is sometimes taken as an indication that a person is overweight. Use this criterion and the data in the previous table to complete this two-way frequency table.

Weight rating	Gender	
	male	female
overweight		
not overweight		
Total	17	21

[2 marks]

d. Does the data support the contention that, for this sample, weight rating is associated with gender? Justify your answer by quoting appropriate percentages.

[2 marks]

e. The parallel boxplots shown here have been constructed to compare the distribution of *BMI* for males and females in this sample.

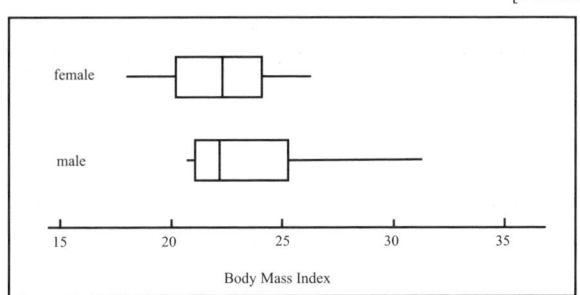

i. Use the parallel boxplots to identify and name two **similar** properties of the *BMI* distributions for males and females.

ii. Use the information in the table in part **a.** to determine the mean *BMI* for the males in this sample. Write your answer correct to one decimal place.

Mean *BMI* for males = ☐

iii. The median *BMI* for males is 22.5. Of the **mean** or **median**, which measure gives a better indication of the typical *BMI* for males? Explain your answer.

[2 + 1 + 2 = 5 marks]
[VCAA 2004 FM]

Question 274

Cars depreciate in value over time. The table gives the average value of a car (of the same brand and model) at different ages.

age (years)	1	2	3	4	5	6	7	8	9
value (dollars)	18 100	15 050	13 900	11 900	10 400	9600	8900	8500	8400

C1. Core – extended responses

a. The data is to be used to build a mathematical relationship that will enable the average value of this brand and model of car to be predicted from its age. In this situation, the **dependent** variable is ☐ [1 mark]

The scatterplot here is constructed from the data and a least squares regression line is fitted as shown.

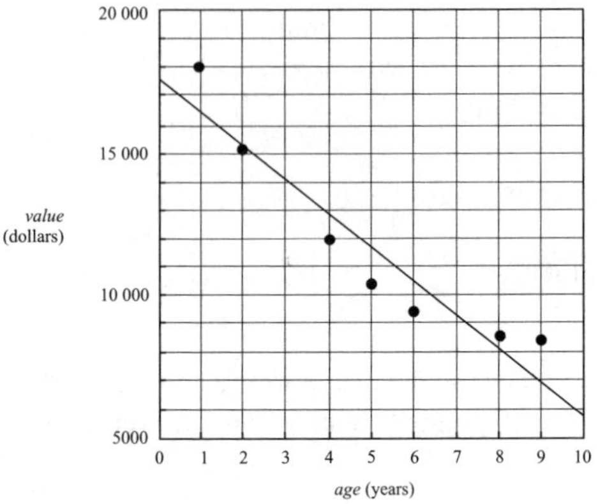

b. The coefficient of determination for this data is 0.9058.

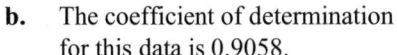

  i. Find the value of the correlation coefficient correct to three decimal places.

  ii. Write down the percentage of the variation in the value of a car that can be accounted for by the variation in its age.

  percentage = ☐

  [1 + 1 = 2 marks]

c. Using the line shown in the scatterplot, or otherwise, determine the equation of the least squares regression line. Write the coefficients correct to the nearest hundred.

value = ☐ + ☐ × age

[2 marks]

The scatterplot suggests that the constructed relationship may be nonlinear. To investigate this idea, a residual plot is constructed. It is incomplete.

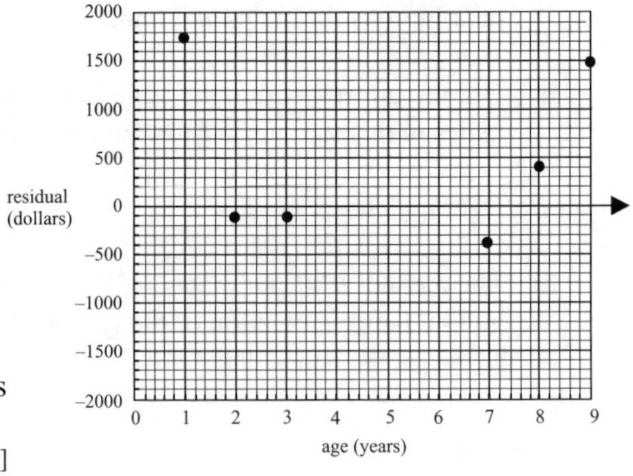

d. Using the information in the scatterplot, or otherwise, complete the residual plot 1 by marking in the missing residual values for cars aged 4, 5 and 6 years.

[2 marks]

e. When complete, does the residual plot suggest that a nonlinear relationship will provide a better fit for the data? Justify your response.

[1 mark]

The scatterplot indicates that a logarithmic transformation of the horizontal (age) axis may linearise the data. The original data has been reproduced in the next table. An extra row has been added for the transformed variable, log(age). The table is incomplete.

C1. Core – extended responses

age (years)	1	2	3	4	5	6	7	8	9
log (age)	0	0.30	0.48	0.60	0.70	0.78	0.85	0.90	
value ($)	18 100	15 050	13 900	11 900	10 400	9600	8900	8500	8400

f. Complete the table above. Write your answer correct to two decimal places.
[1 mark]

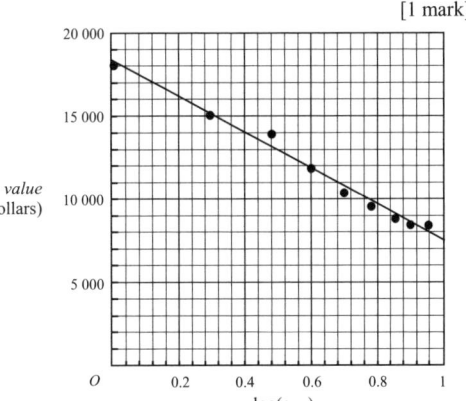

g. In the scatterplot, value is plotted against log(age). A least squares regression line fitted to the transformed data is also drawn. Use the information in this scatterplot to describe the relationship between value and log(age) in terms of **direction**, **form** and **strength**.
[3 marks]

h. The equation of this least squares regression line is
value = 18 300 – 10 800 × log(age).
Use this equation to predict the value of a car that is three years old. Write your answer correct to the nearest hundred dollars.
[1 mark]

The residual plot for this relationship is shown in this residual plot.

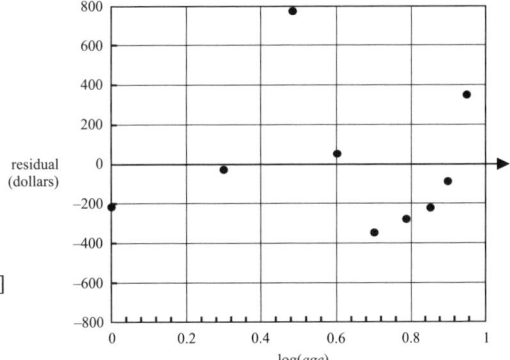

i. This residual plot suggests that the log(age) transformation has been successful in linearising the data. What feature of this residual plot shows that it has been successful?
[1 mark]

j. A transformation applied to the value axis can also linearise the original data displayed in the scatterplot. Suggest a suitable transformation and explain why it will work.
[1 mark]
[VCAA 2005 FM]

Question 275

The table on the next page shows the heights (in cm) of three groups of randomly chosen boys aged 18 months, 27 months and 36 months respectively.

89

C1. Core – extended responses

18 months	height (cm) 27 months	36 months
76.0	82.0	88.0
78.5	83.1	88.8
78.6	84.0	90.0
80.0	86.8	92.3
80.5	87.2	93.0
81.2	87.6	94.1
82.8	88.3	94.2
83.2	90.7	95.8
83.4	91.0	96.9
83.7	92.3	97.1
85.8	92.5	97.8
86.6	93.1	99.2
87.3	94.8	100.6
89.8	97.2	103.8

a. Complete the following table by calculating the standard deviation of the heights of the **18-month-old** boys. Write your answer correct to one decimal place.

age	18 month	27 months	36 months
mean	82.7	89.3	95.1
standard deviation		4.5	4.5

A **27-month-old** boy has a height of 83.1 cm.

b. Calculate his standardised height (z score) relative to this sample of 27-month-old boys. Write your answer correct to one decimal place.

[1 mark]

The heights of the **36-month-old** boys are normally distributed. A 36-month-old boy has a standardised height of 2.

c. Approximately what percentage of 36-month-old boys will be shorter than this child?

percentage =

[1 mark]

C1. Core – extended responses

Using the data from the first table, boxplots have been constructed to display the distributions of heights of 36-month-old and 27-month-old boys as shown below.

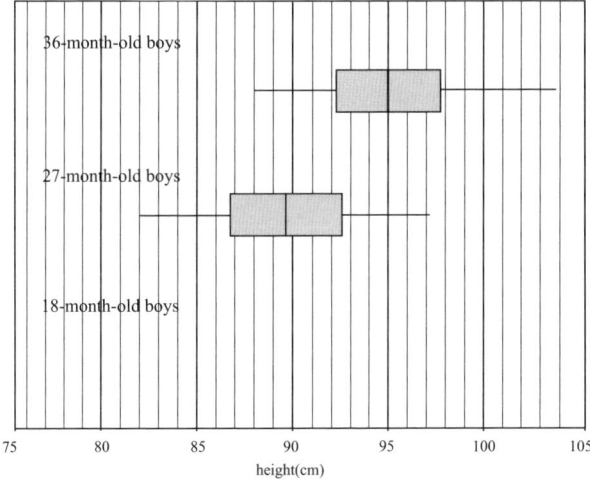

d. Complete the display by constructing and drawing a boxplot that shows the distribution of heights for the **18-month-old** boys.

[2 marks]

e. Use the appropriate boxplot to determine the median height (in centimetres) of the **27-month-old** boys.

[1 mark]

The three parallel boxplots suggest that *height* and *age* (18 months, 27 months, 36 months) are **positively** related.

f. Explain why, giving reference to an appropriate statistic.

[1 mark]
[VCAA 2006 FM]

Question 276

The heights (in cm) and ages (in months) of a different random sample of 15 boys have been plotted in the scatterplot shown here. The least squares regression line has been fitted to the data.

The equation of the least squares regression line is:
height = 75.4 + 0.53 × *age*.
The correlation coefficient is:
r = 0.7541

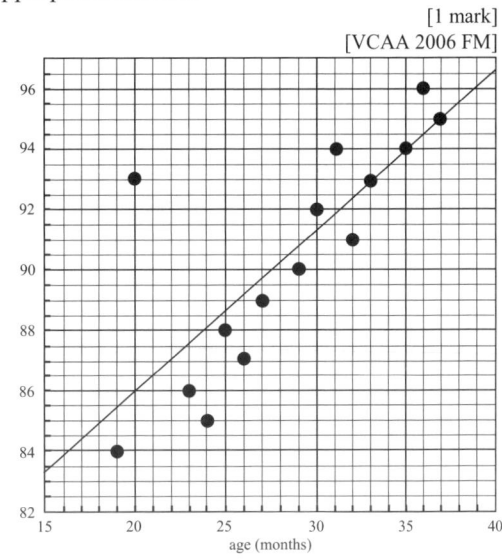

C1. Core – extended responses

a. Complete the following sentence: On average, the height of a boy

   increases by ☐ cm for each one-month increase in age.

   [1 mark]

b. i. Evaluate the coefficient of determination. Write your answer, as a percentage, correct to one decimal place.

   ii. Interpret the coefficient of determination in terms of the variables height and age.

   [1 + 1 = 2 marks]

Question 277

The heights (in cm) and ages (in months) of the 15 boys are shown in the scatterplot below.

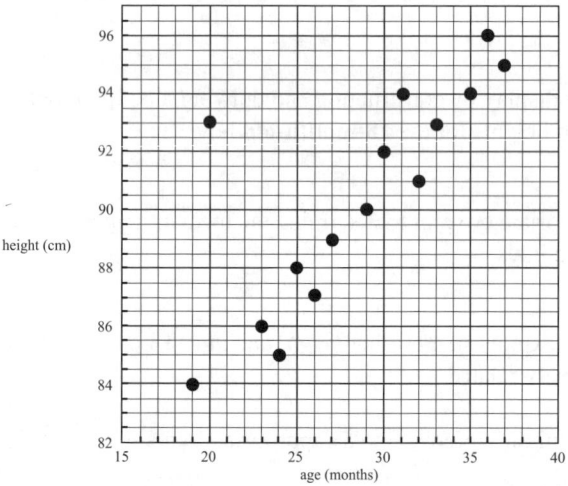

a. Fit a **three median** line to the scatterplot. **Circle** the three points you used to determine this three median line.

   [2 marks]

b. Determine the equation of the three median line. Write the equation in terms of the variables *height* and *age* and give the slope and intercept correct to one decimal place.

   [2 marks]

c. Explain why the three median line might model the relationship between *height* and *age* better than the least squares regression line.

   [1 mark]
   [VCAA 2006 FM]

C1. Core – extended responses

Question 278

The histogram on the right shows the distribution of mean yearly rainfall (in mm) for Australia over 103 years.

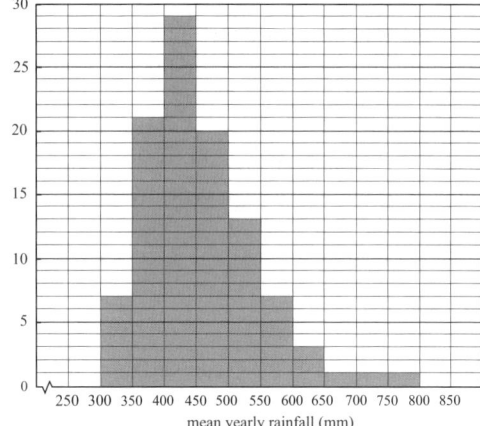

Data source: ABS 2007

a. Describe the shape of the histogram.

[2 marks]

b. Use the histogram to determine

  i. the number of years in which the mean yearly rainfall was 500 mm or more.

  ii. the percentage of years in which the mean yearly rainfall was between 500 mm and 600 mm. Write your answer correct to one decimal place.

[1 + 1 = 2 marks]
[VCAA 2007 FM]

Question 279

The mean surface temperature (in °C) of Australia for the period 1960 to 2005 is displayed in the time series plot shown here.

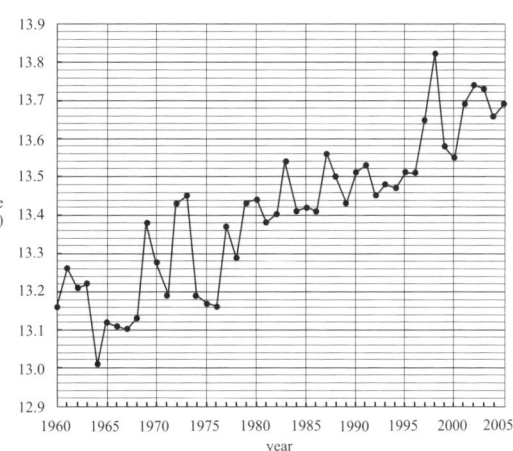

a. In what year was the lowest mean surface temperature recorded?

[1 mark]

The least squares method is used to fit a trend line to the time series plot.

b. The equation of this trend line is found to be
mean surface temperature $= -12.361 + 0.013 \times year$

  i. Use the trend line to predict the mean surface temperature (in °C) for 2010. Write your answer correct to two decimal places.

C1. Core – extended responses

The actual mean surface temperature in the year 2000 was 13.55°C.

    ii. Determine the residual value (in °C) when the trend line is used to predict the mean surface temperature for this year. Write your answer correct to two decimal places.

    iii. By how many degrees does the trend line predict Australia's mean surface temperature will rise each year? Write your answer correct to three decimal places.

[1+1+1 = 3 marks]
[VCAA 2007 FM]

Question 280

The table below displays the mean surface temperature (in °C) and the mean duration of warm spell (in days) in Australia for 13 years selected at random from the period 1960 to 2005.

Mean surface temperature (°C)	Mean duration of warm spell (days)
13.2	21.4
13.3	16.3
13.3	27.6
13.4	32.6
13.4	28.7
13.5	30.9
13.5	45.9
13.5	35.5
13.6	40.6
13.7	42.8
13.7	49.9
13.7	55.8
13.8	**53.1**

This data set has been used to construct the following scatterplot. The scatterplot is incomplete.

    a. Complete the following scatterplot by plotting the bold data values given in the table above. Mark the point with a cross (×).

[1 mark]

C1. Core – extended responses

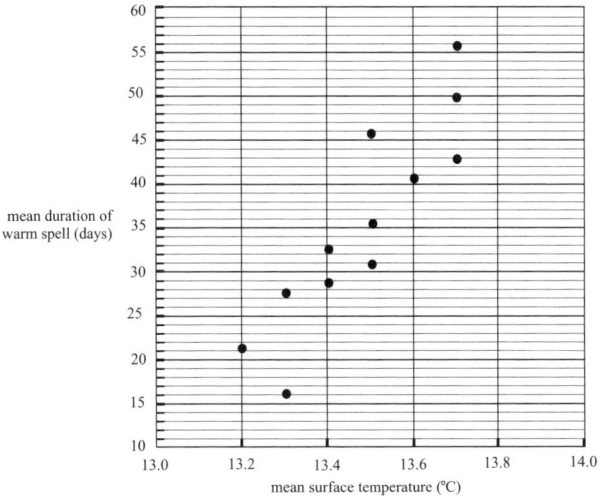

b. Mean surface temperature is the independent variable.

  i. Determine the equation of the least squares regression line for this set of data. Write the equation in terms of the variables mean duration of warm spell and mean surface temperature. Write the values of the coefficients correct to one decimal place.

  ii. Plot the least squares regression line on the scatterplot in part **a**.

  [2 + 1 = 3 marks]

The residual plot shown here was constructed to test the assumption of linearity for the relationship between the variables *mean duration of warm spell* and the *mean surface temperature*.

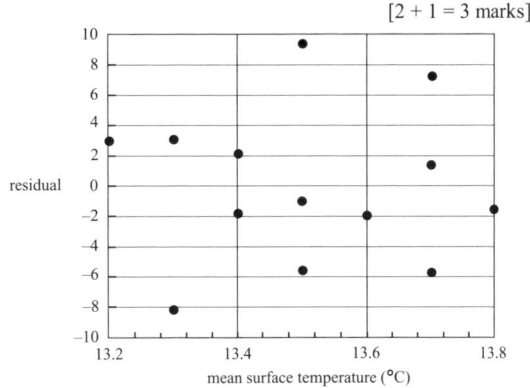

c. Explain why this residual plot supports the assumption of linearity for this relationship.

  [1 mark]

d. Write down the percentage of variation in the mean duration of a warm spell that is explained by the variation in mean surface temperature. Write your answer correct to the nearest per cent.

  [1 mark]

e. Describe the relationship between the mean duration of a warm spell and the mean surface temperature in terms of strength, direction and form.

  [2 marks]
  [VCAA 2007 FM]

C2. Number patterns – extended responses

Question 281

On the O'Callaghans' farm there is only 10 000 litres of water left in their tank. They need to buy more water which is delivered by a water tanker. The water is pumped from the water tanker into the tank at a constant rate of 800 litres per minute.

a. How many litres of water are in the tank one minute after pumping starts?

   ☐ litres
   [1 mark]

b. How many litres of water are in the tank five minutes after pumping starts?

   [1 mark]

c. If t_n litres denotes the amount of water in the tank n minutes after pumping starts, we can write $t_n = a + bn$. Determine the values of a and b.

   $a =$ ☐ $b =$ ☐

   [2 marks]

d. The water tanker has a capacity of 20 000 litres. How long does it take to empty a full tanker at a rate of 800 litres per minute?

   [1 mark]
   [VCAA 2003 FM]

Question 282

With a full tank, the O'Callaghans tend to use more water. As the level of water falls, they become more careful with their water usage. Each week they use 5% of the water that was in the tank at the start of that week. There are 30 000 litres of water in the tank at the **start** of the first week.

a. How many litres of water are in the tank at the **end** of the first week?

   [1 mark]

b. How many litres of water are in the tank at the **end** of the fourth week? Write your answer correct to the nearest litre.

   [1 mark]

c. At the **end** of which week would the amount of water in the tank **first** be less than 10 000 litres?

   [2 marks]
   [VCAA 2003 FM]

C2. Number patterns – extended responses

Question 283

The O'Callaghans use river water to water their crops. As the weather becomes warmer they will use more river water. The amount of river water they use weekly increases in a geometric sequence. In the first week of summer they use 6000 litres of river water. Thereafter their water usage increases by 10% of the previous week's usage.

a. What is the common ratio, r, for this sequence of weekly river water usage?

[1 mark]

b. The O'Callaghans follow this pattern of usage for five weeks. How many litres of river water, in total, will they have used in these five weeks? Write your answer correct to the nearest litre.

[2 marks]

The Gerbers live on a neighbouring property. If G_n denotes the amount of river water used by the Gerbers in the nth week of summer, then the sequence $G_1, G_2, \ldots$ follows the difference equation

$$G_{n+1} = 0.98G_n + 100, \text{ where } G_1 = 8000$$

c. How many litres of river water will the Gerbers use in the third week of summer? Write your answer correct to the nearest litre.

[1 mark]

d. In which week of summer will the O'Callaghans first use more river water for that week than the Gerbers?

[2 marks]
[VCAA 2003 FM]

Question 284

Australian Heating is a company that produces heating systems. The number of heating systems produced annually is modelled by an increasing geometric sequence. The number of heating systems produced in each of the first three years is shown in the following table. Annual production of heating systems:

Year	1	2	3
Number of heating systems produced	2000	2200	2420

a. Show that the common ratio, r, of this geometric sequence is 1.1

[1 mark]

b. What is the annual percentage increase in the number of heating systems produced each year?

[____] %

[1 mark]

C2. Number patterns – extended responses

c. How many heating systems will be produced in year 5? Write your answer correct to the nearest whole number.

[1 mark]

d. The number of heating systems produced annually continues to follow this pattern. In total, how many heating systems will they produce in the first ten years of operation?

[1 mark]

e. The geometric sequence in the table in part **a**. can also be generated by a difference equation of the form $P_{n+1} = bP_n + c$ where $P_1 = 2000$ and P_n is the number of heating systems produced in the nth year. Determine the values of b and c.

$b =$ ☐ $c =$ ☐

[2 marks]
[VCAA 2004 FM]

Question 285

The purchase and installation of a basic heating system with **five** outlets costs $3500. Each additional outlet costs an extra $80.

a. Determine the cost of installing a heating system with **eight** outlets.

[1 mark]

b. A customer has $4400 to spend on a heating system and outlets. Determine the greatest number of outlets that can be bought with this heating system.

[2 marks]

c. Australian Heating recommends that a house with 20 squares of living area should have 12 heating outlets. Using this recommended ratio, determine the **cost** of installing a heating system for a house having 35 squares of living area.

[2 marks]
[VCAA 2004 FM]

Question 286

The number, S_n, of heating systems sold in the nth year is generated by the difference equation $S_n = 1.2S_{n-1} - 200$ where $n \leq 5$ and $S_3 = 2224$.

a. Use the difference equation to determine how many heating systems were sold in the first year.

[2 marks]

b. What percentage of heating systems produced during the first three years was sold within the three years? Write your answer correct to one decimal place.

[2 marks]
[VCAA 2004 FM]

C2. Number patterns – extended responses

Question 287

Noel is an enthusiastic runner. He decides to follow a training program over several weeks. The distance (in kilometres) that Noel runs each day forms an arithmetic sequence. The distances for the first three days of this training program are shown in the table below.

Day	1	2	3
Distance (km)	4.2	4.5	4.8

a. What distance (in kilometres) will Noel run on Day 5?

[1 mark]

b. On which day will Noel first run more than 7 km?

[1 mark]

c. Determine the total distance that Noel runs from Day 3 to Day 12 inclusive. Write your answer in kilometres, correct to one decimal place.

[2 marks]

d. An expression for the *n*th term of this sequence can be written as $t_n = 0.3n + b$. Determine the value of b.

[1 mark]
[VCAA 2005 FM]

Question 288

Ray enjoys walking. The distance (in kilometres) that Ray walks each day forms a sequence. The distances for the first three days of his walking program are shown in the table below.

Day	1	2	3
Distance (km)	3	3.5	4.05

a. Show that this sequence is neither arithmetic nor geometric.

[2 marks]

b. A difference equation that generates the terms of this sequence is
$w_{n+1} = 1.1w_n + c$ where $w_1 = 3$.
What is the value of c?

[1 mark]

c. What distance will Ray walk on Day 5? Write your answer in kilometres, correct to two decimal places.

[1 mark]
[VCAA 2005 FM]

C2. Number patterns – extended responses

Question 289

Catherine likes both running and walking. She commences a two week training program. The distance (in kilometres) that Catherine travels each day of the training program is always 5% greater than that of the previous day. The distance travelled in the first and third days of her training program are shown in the following table:

Day	1	2	3
Distance (km)	10		11.025

a. Complete the table above.

[1 mark]

b. Calculate the total distance that Catherine travels in 14 days. Write your answer correct to the nearest kilometre.

[1 mark]

c. Catherine's training each day consists of both running and walking.

    i. On Day 1, Catherine travelled 10 km. The ratio of her running distance to her walking distance was 3:2. Determine the distance that Catherine walked on Day 1.

    ii. On Day 3, Catherine ran the first 25% of the total distance. Determine the ratio of her running distance to her walking distance on Day 3.

[1 + 1 = 2 marks]

d. Catherine will begin a new training program at the start of next month. On the first day of this new program she will travel 5 km. Each day thereafter she will travel 90% of the previous distance plus an additional 1.2 km.

    i. Write a difference equation that **specifies** the distance Catherine will travel on the nth day.

Catherine follows this training program for many months.

    ii. Explain why she will never travel more than 12 km in any one day.

[1 + 1 = 2 marks]
[VCAA 2005 FM]

Question 290

It is estimated that the trees in an orchard contain 48 000 kg of fruit. Each day 3000 kg of fruit is picked from the trees.

a. How many kilograms of fruit remain on the trees at the end of the second day?

[1 mark]

b. The number of kilograms of fruit, $F\sim$, remaining on the trees at the end of the nth day can be written as $F = 48\,000 + d \times n$. Find the value of d.

[1 mark]

C2. Number patterns – extended responses

c. How many days, in total, will it take to pick all the fruit from the trees?
[1 mark]
[VCAA 2006 FM]

Question 291

A new type of fruit tree is planted in the orchard. In the first month after planting, the gardeners worked 625 hours. In the second and third months after planting, the gardeners worked 500 hours and 400 hours respectively. Suppose this decreasing pattern of work continues. The number of hours the gardeners work each month follows a geometric sequence.

a. Show that the common ratio of this sequence is $r = 0.8$
[1 mark]

b. Determine the number of hours the gardeners will work in the fifth month.
[1 mark]

c. Write an expression that gives the number of hours, H_n, the gardeners will work in the nth month after planting.
[1 mark]

d. How many more hours will the gardeners work in the sixth month than in the seventh month? Write your answer correct to the nearest hour.
[1 mark]

e. In which month will the gardeners first work less than 100 hours?
[1 mark]

In the first three months after planting, the gardeners worked a total of 1525 hours.

f. How many hours, in total, will the gardeners work in the next nine months? Write your answer correct to the nearest hour.
[2 marks]
[VCAA 2006 FM]

Question 292

The water used in the orchard is stored in a tank. Each afternoon, 10% of the volume of water in the tank is used. Each evening, 2000 litres of water is added to the tank. This pattern continues each day. The volume of water, V_n, in the tank on the morning of the nth day is modelled by the difference equation

$$V_{n+1} = rV_n + d \text{ where } V_1 = 45\,000 \text{ litres.}$$

a. Find r and d.
[2 marks]

C2. Number patterns – extended responses

b. Determine how many litres of water will be in the tank on the morning of the fourth day.

[1 mark]

c. On the morning of which day will the volume of water in the tank first be below 30 000 litres?

[1 mark]

d. In the long term, how many litres of water will be in the tank each morning? Write your answer correct to the nearest litre.

[1 mark]
[VCAA 2006 FM]

Question 293

Maria intends to follow a healthy eating plan. She will reduce her daily kilojoule intake by a constant amount each day over a period of 14 days.

The graph here shows Maria's kilojoule intake for the first five days.

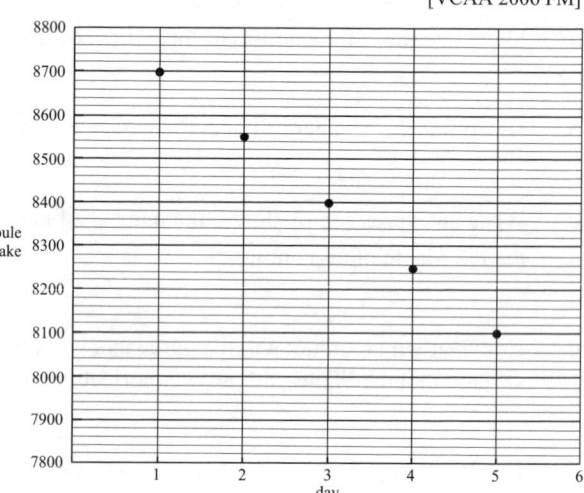

a. What was Maria's kilojoule intake on day 1?

[1 mark]

b. Determine Maria's kilojoule intake on day 6.

[1 mark]

c. Maria's kilojoule intake on the nth day is given by the equation $K_n = a - 150 \times n$. Determine the value of a.

[1 mark]

d. On which day will Maria's daily kilojoule intake be 6750?

[1 mark]
[VCAA 2007 FM]

Question 294

Maria's brother, Rupert, believes he will benefit by reducing his daily kilojoule intake. His kilojoule intake over 14 days will follow a geometric sequence with a common ratio of 0.95. On day 1, Rupert's kilojoule intake was 12 000.

a. By what percentage is Rupert's kilojoule intake reduced each day?

[[1 mark]

b. Determine Rupert's kilojoule intake on day 3.

[1 mark]

C2. Number patterns – extended responses

c. Write an equation that gives Rupert's kilojoule intake R_n on the nth day.

[1 mark]

d. Find the difference between Rupert's kilojoule intake on day 9 and day 10. Write your answer correct to the nearest kilojoule.

[1 mark]

e. Determine Rupert's total kilojoule intake from day 8 to day 14 inclusive. Write your answer correct to the nearest kilojoule.

[2 marks]
[VCAA 2007 FM]

Question 295

Maria decides to improve her fitness level by cycling each day.
The time in minutes M_n that Maria cycles on the nth day is modelled by the difference equation
$$M_{n+1} = 0.75 M_n + 8 \quad \text{where} \quad M_2 = 20$$

a. For how many minutes will Maria cycle on day 4?

[1 mark]

b. Show that the time Maria cycles each day does not follow an arithmetic or a geometric sequence.

[1 mark]

c. For how many minutes will Maria cycle on day 1?

[2 marks]
[VCAA 2007 FM]

Question 296

Rupert decides to include both swimming and running in his exercise plan. On day 1, Rupert swims 100 m and runs 500 m.
Each day he will increase the distance he swims and the distance he runs.
His swimming distance will increase by 50 m each day.
His running distance will increase by 2% of the distance he ran on the previous day.
On which day will the distance Rupert swims first be greater than the distance he runs?

[2 marks]
[VCAA 2007 FM]

C3. Geometry and trigonometry – extended responses

The following information is for questions 297, 298 and 299.

Hudson takes a kite out and flies it on a 20 metre string that is connected to the kite as shown. At a particular instant when the kite is in the air, the horizontal distance of the kite to his hand is 8 metres. The angle the kite string makes with the horizontal is θ.

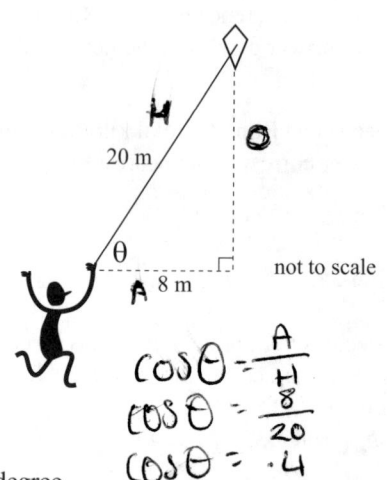

not to scale

Question 297

a. Calculate the angle θ correct to the nearest degree.

[2 marks]

b. The kite begins to rise while the length of the string remains 20 metres. At a particular instant, the horizontal distance from his hand to the kite is 5 metres. Find how high the kite is above his hand, at that instant, correct to the nearest metre.

[2 marks]

Hudson's daughter Dayani is standing east of her father at *D* when the kite crashes into the ground. The crash site, *C*, is on a bearing of 310° and a distance of 18 m from Dayani. Hudson is 20 metres from the crash site at *H* as shown.

Question 298

a. For △*HDC*, find the size of ∠*HDC* in degrees.

[1 mark]

b. Hence or otherwise, find the size of ∠*CHD* in degrees. Give your answer correct to two decimal places.

[2 marks]

This diagram shows the kite, *ABCD*. The length of side *AB* is 40 centimetres and *BC* is 80 centimetres. The size of ∠*ABC* is 125°.

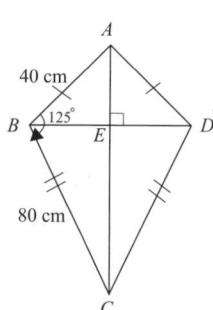

104

C3. Geometry and trigonometry – extended responses

Question 299

The frame of the kite is made from two supporting pieces of lightweight wood, AC and BD, which cross at right angles.

a. Find the length AC, correct to the nearest centimetre.

[2 marks]

b. The kite is covered with a lightweight fabric which is cut exactly to fit the kite shape. Assuming there is no overlapping of the fabric, show that the area of fabric needed to make the kite is 2621 cm² to the nearest square centimetre.

[2 marks]

c. For the kite $ABCD$, determine the size of $\angle ABE$ correct to the nearest degree.

[3 marks]

d. Hudson decides to make a larger kite, $PQRS$, with corners labelled as shown. For the two kites, the ratio $AD:PS = CD:RS = 2:3$. Given that the area of the fabric for kite $ABCD$ is 2621 cm², determine the area of fabric needed for the kite $PQRS$ correct to the nearest square centimetre.

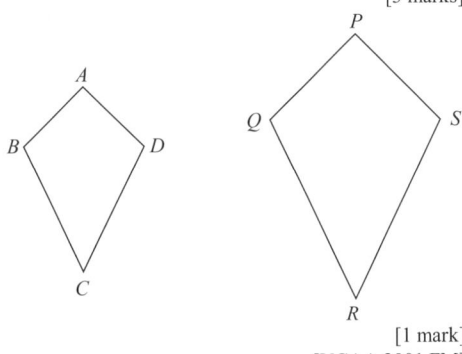

[1 mark]
[VCAA 2001 FM]

Question 300

The diagram here shows a camping ground by the sea. The boundary PQ is 1203 metres long and runs beside an east–west road. The boundary PS is 1048 metres long and the bearing of S from P is 015° true. The boundary QR is 951 metres long and the angle PQR is 80°. The fourth boundary of the camping ground is along a cliff edge by the sea.

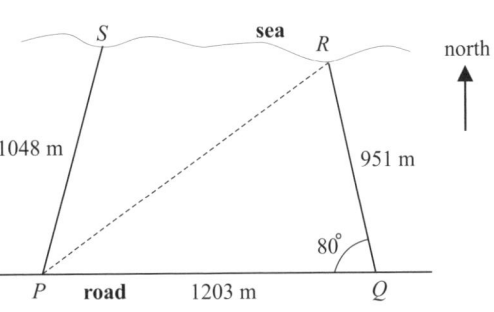

a. Find the area of triangle PQR in square metres, correct to the nearest square metre.

[2 marks]

b. Find the distance from P to R, correct to the nearest metre.

[2 marks]

c. Find the bearing of R from P, correct to the nearest degree.

[2 marks]

C3. Geometry and trigonometry – extended responses

d. Find the area of triangle *PSR* in square metres, correct to the nearest 10 square metres.

[2 marks]

e. Hence find the approximate area of the camping ground. Give your answer correct to the nearest 10 square metres.

[1 mark]
[VCAA 2002 FM]

Question 301

Jan is landscaping her garden. A piece of shade cloth, *ABC*, has dimensions as shown here.

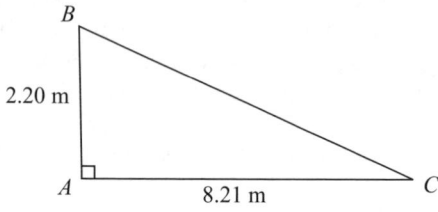

a. Determine the length, *BC*, in metres. Write your answer correct to two decimal places.

[1 mark]

b. Determine the angle *ACB*. Write your answer correct to the nearest degree.

[1 mark]
[VCAA 2003 FM]

Question 302

A paved area is constructed in the shape of a regular octagon as shown here.

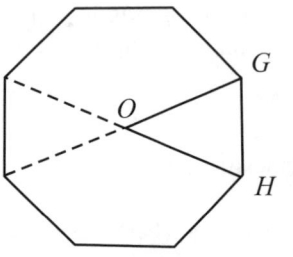

a. By calculation, show that the size of angle *GOH* is 45°, where point *O* is the centre of the octagon.

[1 mark]

b. The length *OG* = *OH* = 2.30 metres. Calculate the area of the octagonal paved area. Write your answer correct to the nearest square metre.

[1 mark]

A square herb garden *EFGH* is surrounded by four regular octagonal paved areas as shown in the diagram.

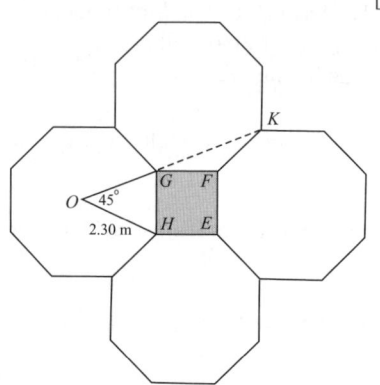

c. Calculate the side length *GH* of the square herb garden. Write your answers in metres, correct to two decimal places.

[1 mark]

106

C3. Geometry and trigonometry – extended responses

d. A straight wooden frame is to be built between points O and K for hanging baskets.

  i. Calculate the length GK. Write your answer in metres, correct to two decimal places.

  ii. Hence calculate the length OK. Write your answer in metres, correct to two decimal places.

  [2 + 1 = 3 marks]

A second piece of shade cloth PQR is also triangular and has dimensions as shown in the daigram.

e. Calculate the length of PR. Write your answer in metres, correct to two decimal places.

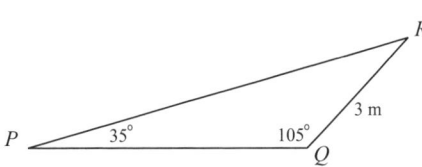

[1 mark]

The second piece of shade cloth PQR is attached to three vertical poles located at X, Y and Z as shown in the diagram. Poles PX and QY are each 3.5 metres long. The horizontal distance YZ is 2.7 metres.

f. Calculate the length of the vertical pole RZ. Write your answer correct to the nearest **centimetre**.

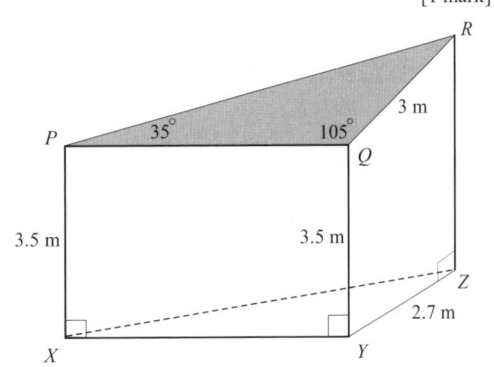

[2 marks]
[VCAA 2003 FM]

Question 303

Jane has soil delivered for her garden. There are two piles of soil, both in the shape of a right cone.

a. The first pile of soil has a base diameter of 1.2 metres and a height of 0.7 metres as shown in the diagram here.

[1 mark]

Calculate the volume of soil in the first pile. Write your answer in cubic metres, correct to two decimal places.

b. The second pile of soil has a base diameter of 2.4 metres and a height of 1.4 metres. What is the ratio of the volume of the first pile to the volume of the second pile?

[1 mark]
[VCAA 2003 FM]

C3. Geometry and trigonometry – extended responses

Question 304

A yacht has two flat triangular sails as shown in the diagram.

The sail *ABC* is in the shape of a right-angled triangle. The height *AC* is 10 metres and the length *AB* is 3.6 metres.

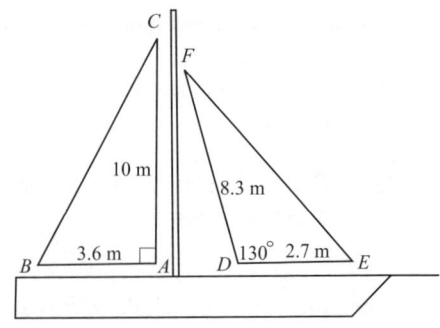

a. Calculate angle *ABC*. Write your answer correct to the nearest degree.

[1 mark]

b. Calculate the length *BC*. Write your answer in metres, correct to one decimal place.

[1 mark]

The sail *DEF* has side lengths *DE* = 2.7 metres and *DF* = 8.3 metres. The angle *EDF* is 130°.

c. Calculate the length *EF*. Write your answer in metres, correct to one decimal place.

[1 mark]

d. Calculate the area of the sail *DEF*. Write your answer in square metres, correct to one decimal place.

[1 mark]
[VCAA 2004 FM]

Question 305

A course for a yacht race is triangular in shape and is marked by three buoys *T*, *U* and *V*.

Starting from buoy *V*, the yachts sail 5.4 kilometres on a bearing of 030° to buoy *T*. They then sail to buoy *U* and back to buoy *V*. The angle *TVU* is 72° and the angle *TUV* is 48°.

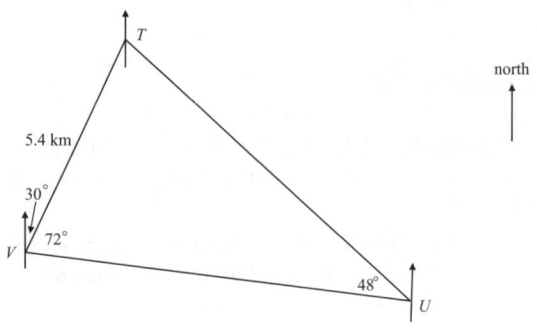

a. Determine the bearing of *V* from *U*.

[2 marks]

b. Determine the distance *TU*. Write your answer in kilometres, correct to one decimal place.

[1 mark]

C3. Geometry and trigonometry – extended responses

c. Determine the shortest distance to complete the race. Write your answer in kilometres, correct to one decimal place.

[2 marks]
[VCAA 2004 FM]

Question 306

A navigational marker *XYZ* is in the shape of an equilateral triangle with side length of one metre. It is located in the vicinity of the yacht race.

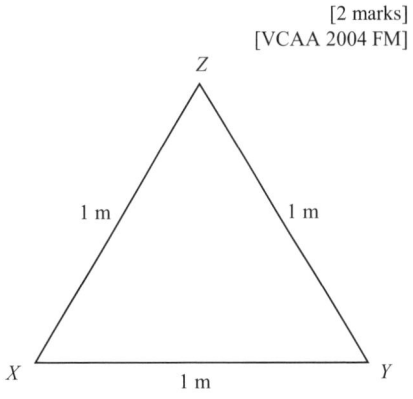

a. Write down the size of angle *XYZ*.

angle *XYZ* =

[1 mark]

Point *O* is the centroid (centre) of the triangle. Points *M* and *N* are the midpoints of sides *XZ* and *YZ* respectively.

b. Calculate the shortest distance from point *O* to side *XY*. Write your answer in metres, correct to three decimal places.

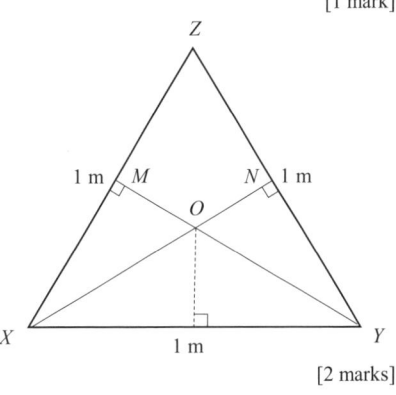

[2 marks]

A piece of reflective material in the shape of a circle is attached to the centre of the navigational marker at the centroid *O*. The ratio of the area of the shaded region of the navigational marker *XYZ* to the area of the reflective material is 2:1.

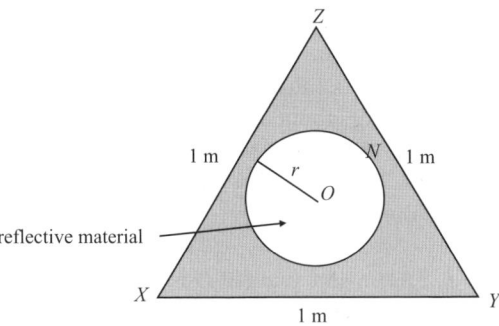

c. Determine the radius, *r*, of the circle. Write your answer in metres, correct to three decimal places.

[3 marks]
[VCAA 2004 FM]

C3. Geometry and trigonometry – extended responses

Question 307

The contour map shows a region in country Victoria. It has contours drawn at intervals of 50 metres. There is a camping ground at G, a hut at H and a water source at W.
scale 1:40 000

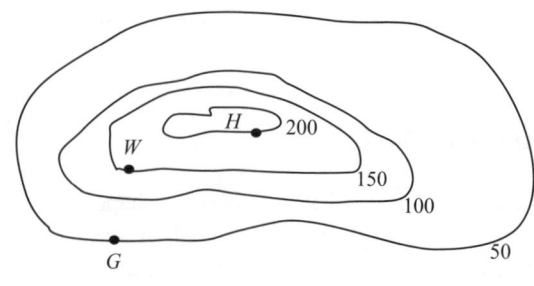

scale 1:40 000

a. What is the difference in height (in metres) between hut H and the water source W?

[1 mark]

b. Camping ground G obtains its water from W. The horizontal distance between G and W is 500 m. Find the shortest length of water pipe that can connect G with W. Write your answer correct to the nearest metre.

[1 mark]

c. The average slope between camping ground G and hut H is 0.12. Find the horizontal distance (in metres) between G and H.

[1 mark]

d. The scale used on this contour map is 1:40 000. Determine the length of a line (in centimetres) on the contour map that represents a distance of 2 km.

[1 mark]
[VCAA 2005 FM]

Question 308

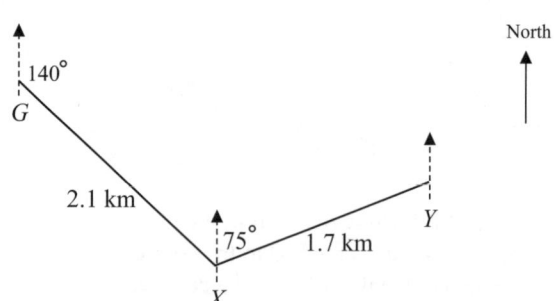

The land near the camping ground is flat and suitable for orienteering. Checkpoint X is situated 2.1 km from camping ground G on a bearing of 140°. Checkpoint Y is situated 1.7 km from checkpoint X on a bearing of 075°.

a. **i.** How far south of G is checkpoint X? Write your answer in kilometres, correct to one decimal place.

 ii. How far south of G is checkpoint Y? Write your answer in kilometres, correct to one decimal place.

[1 + 2 = 3 marks]

b. Determine the size of angle GXY.

[1 mark]

C3. Geometry and trigonometry – extended responses

c. Calculate the distance *GY*. Write your answer in kilometres, correct to one decimal place.

[1 mark]

d. Determine the bearing of checkpoint *Y* from camping ground *G*. Write your answer correct to the nearest degree.

[2 marks]
[VCAA 2005 FM]

Question 309

The cross-section, *ABC*, of an A-frame hut near the camping ground is an isosceles triangle as shown here. Angle *ACB* is 50°. *AB* is 5 metres in length and is the floor line. MN is 2 metres in length and is the ceiling line inside the hut. *AB* and *MN* are parallel.

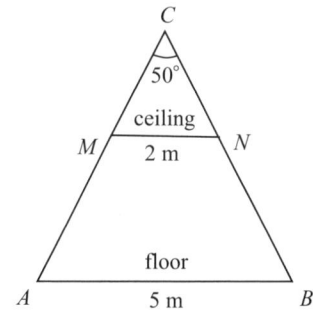

a. Calculate the vertical height of the ceiling, *MN*, above the floor, *AB*. Write your answer in metres, correct to one decimal place.

[2 marks]

b. The hut has the shape of a triangular prism. The space inside the hut above the ceiling is used for storage. The total space inside the hut (including storage) is V m³. What fraction of V is used for storage?

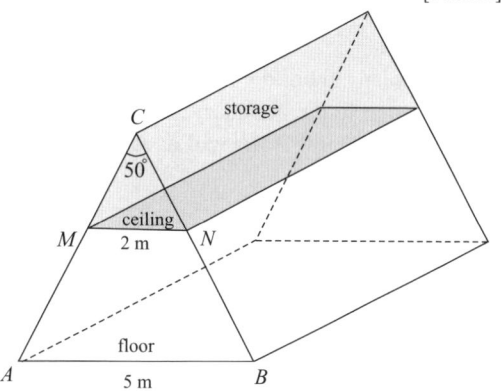

[1 mark]
[VCAA 2005 FM]

Question 310

A farmer owns a flat allotment of land in the shape of triangle *ABC* shown below.

Boundary *AB* is 251 metres. Boundary *AC* is 142 metres. Angle BAC is 45°. A straight track, *XY*, runs perpendicular to the boundary *AC*. Point *Y* is 55 m from *A* along the boundary *AC*.

a. Determine the size of angle *AXY*.

[1 mark]

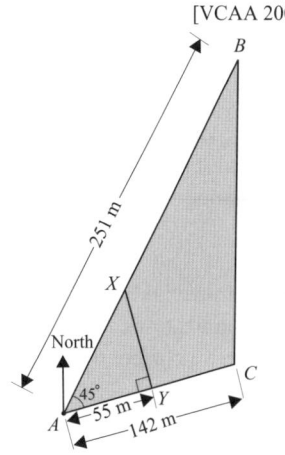

C3. Geometry and trigonometry – extended responses

b. Determine the length of *AX*. Write your answer, in metres, correct to one decimal place.

[1 mark]

c. The bearing of C from A is 078°. Determine the bearing of *B* from *A*.

[1 mark]

d. Determine the shortest distance from *X* to *C*. Write your answer, in metres, correct to one decimal place.

[2 marks]

e. Determine the area of triangle *ABC* correct to the nearest square metre.

[1 mark]

The length of the boundary *BC* is 181 metres (correct to the nearest metre).

f. **i.** Use the cosine rule to show how this length can be found.

ii. Determine the size of angle *ABC*. Write your answer, in degrees, correct to one decimal place.

[1 + 1 = 2 marks]

A farmer plans to build a fence, *MN*, perpendicular to the boundary *AC*. The land enclosed by triangle *AMN* will have an area of 3200 m².

g. Determine the length of the fence *MN*.

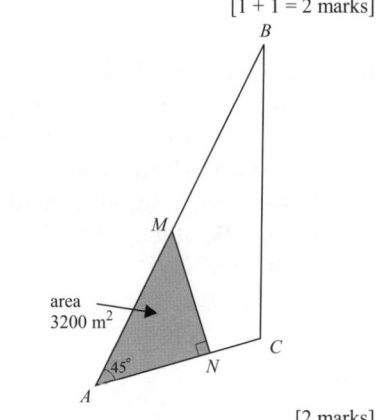

[2 marks]
[VCAA 2006 FM]

Question 311

The allotment of land contains a communications tower, *PQ*. Points *S*, *Q* and *T* are situated on level ground. From *S* the angle of elevation of *P* is 20°. Distance *SQ* is 125 metres. Distance *TQ* is 98 metres.

a. Determine the height, *PQ*, of the communications tower. Write your answer, in metres, correct to one decimal place.

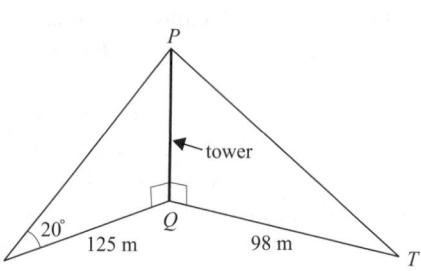

[1 mark]

112

C3. Geometry and trigonometry – extended responses

b. Determine the angle of depression of *T* from *P*. Write your answer, in degrees, correct to one decimal place.

Question 312

A closed cylindrical water tank has external diameter 3.5 metres. The external height of the tank is 2.4 metres. The walls, floor and top of the tank are made of concrete 0.25 cm thick.

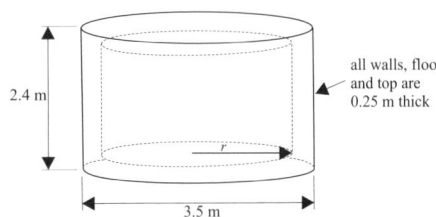

a. What is the internal radius, *r*, of the tank?

[1 mark]

b. Determine the maximum amount of water this tank can hold. Write your answer correct to the nearest cubic metre.

[2 marks]
[VCAA 2006 FM]

Tessa is a student in a woodwork class. The class will construct geometrical solids from a block of wood. Tessa has a piece of wood in the shape of a rectangular prism.
This prism, *ABCDQRST*, shown here, has base length 24 cm, base width 28 cm and height 32 cm.

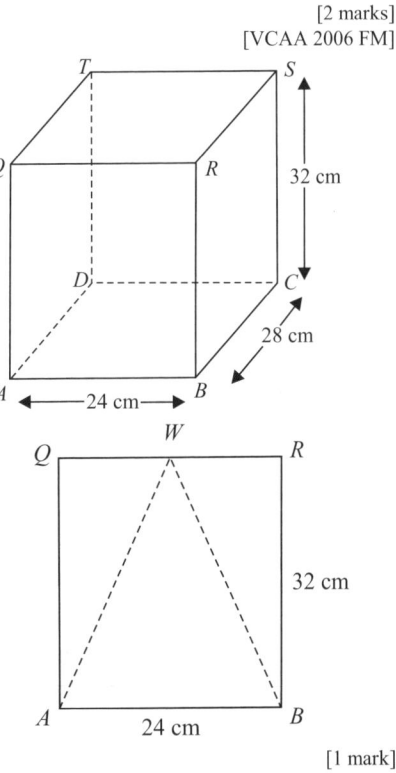

Question 313

On the front face, *ABRQ*, Tessa marks point *W* halfway between *Q* and *R* as shown in the figure here. She then draws line segments *AW* and *BW* as shown here.

a. Determine the length, in cm, of *QW*.

[1 mark]

b. Calculate the angle *WAQ*.
Write your answer in degrees, correct to one decimal place.

[1 mark]

C3. Geometry and trigonometry – extended responses

c. Calculate the angle *AWB* correct to one decimal place.

[1 mark]

d. What fraction of the area of the rectangle *ABRQ* does the area of the triangle *AWB* represent?

[1 mark]
[VCAA 2007 FM]

Question 314

Tessa carves a triangular prism from her block of wood. Using point *V*, halfway between *T* and *S* on the back face, *DCST*, she constructs the triangular prism shown here.

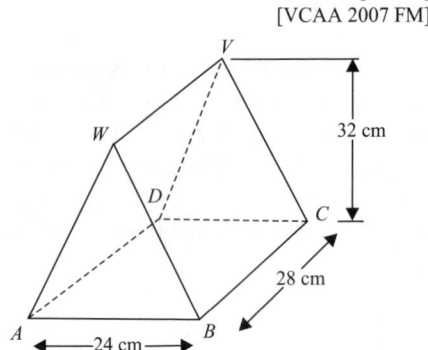

a. Show that, correct to the nearest centimetre, length *AW* is 34 cm.

[2 marks]

b. Using length *AW* as 34 cm, find the total surface area, in cm^2, of the triangular prism *ABCDWV*.

[2 marks]
[VCAA 2007 FM]

Question 315

Tessa's next task is to carve the right rectangular pyramid *ABCDY* shown here. She marks a new point, *Y*, halfway between points *W* and *V* in the diagram in the previous question. She uses point *Y* to construct this pyramid.

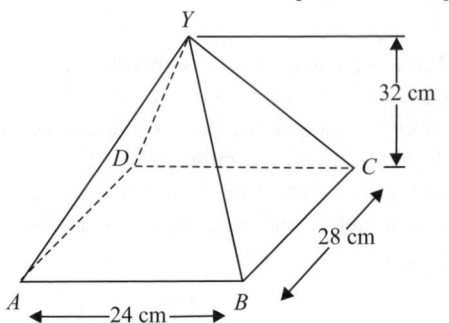

a. Calculate the volume, in cm^3, of the pyramid *ABCDY*.

[1 mark]

b. Show that, correct to the nearest cm, length *AY* is 37 cm.

[2 marks]

c. Using *AY* as 37 cm, demonstrate the use of **Heron's formula** to calculate the area, in cm^2, of the triangular face *YAB*.

[2 marks]
[VCAA 2007 FM]

114

C3. Geometry and trigonometry – extended responses

Question 316

Tessa's final task involves removing the top 24 cm of the height of her pyramid (see the previous question). The shape remaining is shown here. The top surface, *JKLM*, is parallel to the base, *ABCD*.

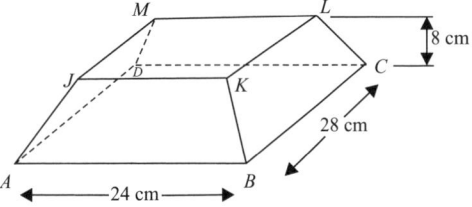

a. What fraction of the height of the pyramid has Tessa removed to produce this shape?

[1 mark]

b. What fraction of the volume of the pyramid remains in this shape?

[2 marks]
[VCAA 2007 FM]

C4. Graphs and relations – extended responses

Question 317

Two bushwalkers, Malinda and Christos, set out to walk from Fishbone Creek to Snake Gully, a distance of 20 km. They start at the same time and follow the same route.

a. Malinda walks at a constant speed of 4 km/h for the entire journey and takes no rest periods. How far does she travel in 1.5 hours?

[1 mark]

b. The distance walked by Malinda from Fishbone Creek, in kilometres, is given by the equation $D_m = 4t$ for $0 \leq t \leq 5$, where t is the time in hours since she began walking. Draw and label the graph of D_m against t on the set of axes below.

[2 marks]

Christos started walking at the same time as Malinda and followed the same route. At the start he walked at a constant speed of 6 km/h. However, after walking at this speed for two hours he developed sore feet. Rather than stopping, he slowed down to a constant speed of 2 km/h for the remainder of the trip.

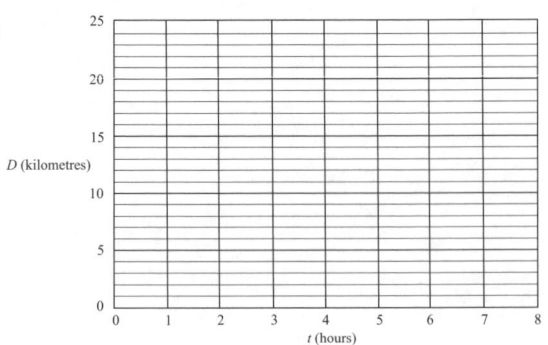

c. Let D_c represent the distance walked by Christos. Draw and label the graph of D_c against t on the set of axes in part **b**.

[2 marks]

d. Malinda eventually catches up to Christos. How many hours after they start walking does this happen?

[1 mark]

e. The equations below give the distance, D_c in kilometres, walked by Christos at any time t hours.

$$D_c = \begin{cases} at & 0 \leq t \leq 2 \text{ hours} \\ bt + h & 2 < t \leq d \text{ hours} \end{cases}$$

Determine the values for a, b, h and d.

[3 marks]
[VCAA 2003 FM]

C4. Graphs and relations – extended responses

Question 318

Malinda began the walk with 2000 millilitres of water in her bottle. It was a hot day and she sipped small amounts of water from her bottle frequently to ensure that she would not dehydrate.

The table below shows the total volume, V, of water, in millilitres, that Malinda had drunk after t hours.

t (hours)	0	2	4	5
t^2 (hours^2)	0			
V (millilitres)	0	300	1200	1875

a. i. Complete the table above.

   ii. On the set of axes below, plot the four points (t^2, V) from the table you have completed.

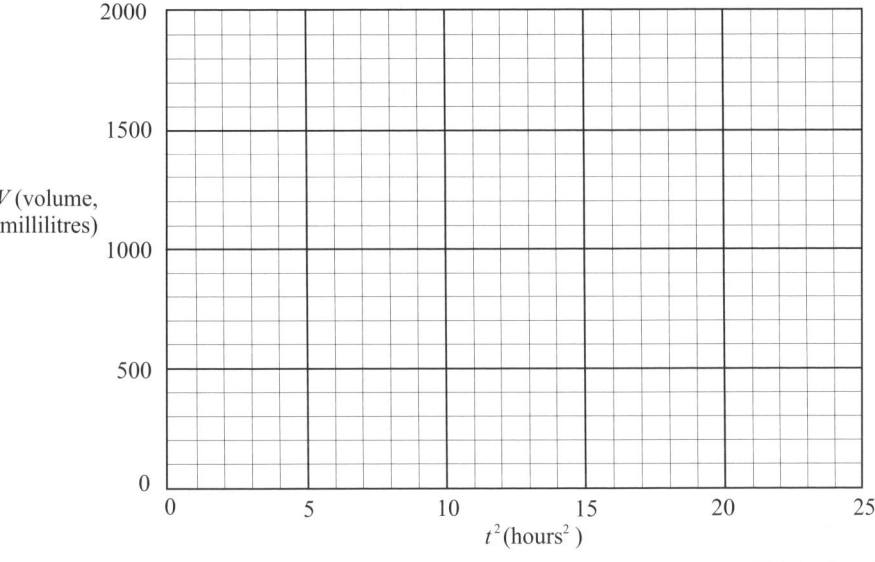

[1 + 1 = 2 marks]

b. Assume that this data is modelled by the relation $V = kt^2$. Use the graph or table to determine the value of k.

[1 mark]

c. Use your answer from **part b** to determine the volume of water Malinda drank during the first three hours. Write your answer correct to the nearest millilitre.

[1 mark]

C4. Graphs and relations – extended responses

d. At some point during the walk, Malinda had drunk half the water in her bottle.

    i. How long after the start of the walk did this happen? Write your answer in hours, correct to two decimal places.

    ii. Determine how far apart Malinda and Christos were at this time. Write your answer in kilometres, correct to one decimal place.

[1 + 1 = 2 marks]
[VCAA 2003 FM]

Question 319

A clothing manufacturer finds that the cost, C dollars, of producing x shirts is given by the equation $C = 8x + 2400$.

a. Determine the cost of producing 400 shirts.

[1 mark]

b. Determine the maximum number of shirts that can be produced for $3000.

[1 mark]

c. Assuming all the shirts are sold, the revenue, R dollars, from the sale of x shirts produced is given by an equation $R = 23x$.
A graph of the revenue equation for $0 \leq x \leq 400$ is drawn on the axes here. On these same axes draw a graph of the cost equation $C = 8x + 2400$ for $0 \leq x \leq 400$.

[2 marks]

d. Determine the number of shirts that need to be produced and sold for the manufacturer to break even.

[1 mark]

e. Given the cost equation is $C = 8x + 2400$ and the revenue equation is $R = 23x$, write an equation for the profit, P dollars, from the production and sale of x shirts.

[2 marks]

f. Calculate the profit from the production and sale of 345 shirts.

[1 mark]

[VCAA 2004 FM]

C4. Graphs and relations – extended responses

Question 320

The manufacturer also produces jackets. They receive an order for 250 jackets. The cost of producing the 250 jackets is $4800. Determine the selling price per jacket to achieve an overall profit of $3000.

[2 marks]
[VCAA 2004 FM]

Question 321

Singlets are produced and sold in larger quantities. The revenue, R_S dollars, generated from the sale of x singlets is given by the equation

$$R_S = \begin{cases} 10x & x \leq 500 \\ 6x + 2000 & x > 500 \end{cases}$$

a. Calculate the revenue, R_S, generated by the sale of 620 singlets.

[1 mark]

b. Sketch a graph of the revenue, R_S, for $0 \leq x \leq 1000$ on the axes here.

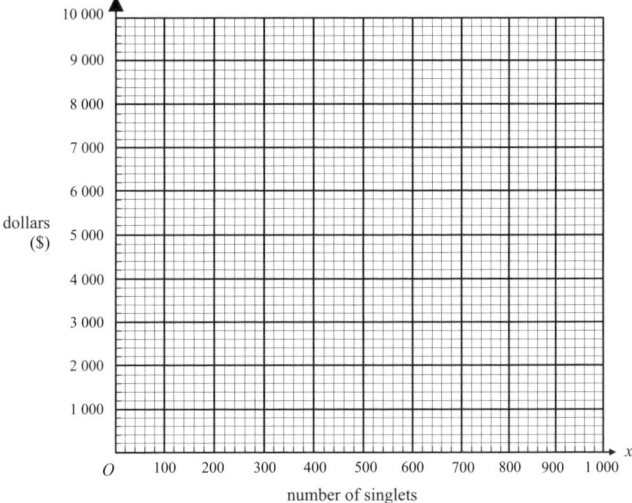

[2 marks]

c. If the cost, C_S dollars, of producing x singlets is $C_S = 4x + 1500$, determine the number of singlets that would need to be produced and sold to obtain a profit of $2000.

[2 marks]
[VCAA 2004 FM]

Question 322

Pepi manages two street stands in the city. One is in Xenon Street and the other is in Yarra Street. These street stands sell newspapers and magazines.
On average, per hour, each stand sells
Xenon Street: 20 newspapers and 2 magazines
Yarra Street: 15 newspapers and 4 magazines.

C4. Graphs and relations – extended responses

In total, at least 900 newspapers and at least 160 magazines must be sold each week.
Let x be the number of hours per week the Xenon Street stand operates
      y be the number of hours per week the Yarra Street stand operates.

This information can be expressed as Inequalities 1 to 4.
Inequality 1: $20x + 15y \geq 900$
Inequality 2: $2x + 4y \geq 160$
Inequality 3: $x \geq 0$
Inequality 4: $y \geq 0$

a. Which line (Line A or Line B) in the graph below forms the boundary of the region defined by Inequality 1: $20x + 15y \geq 900$?

[1 mark]

b. Write down the coordinates of the point of intersection of Line A and Line B in the graph below.

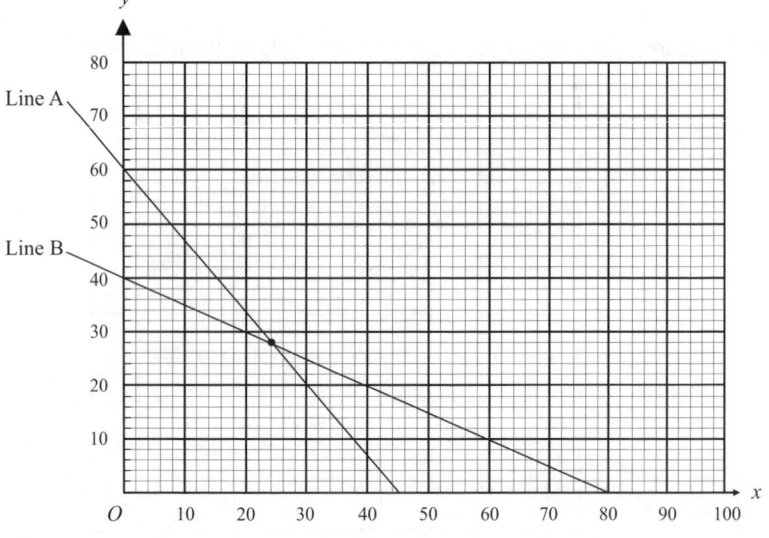

[1 mark]

c. Pepi is allowed to operate the Xenon Street stand for up to 64 hours per week and the Yarra Street stand for up to 52 hours per week. Write two corresponding inequalities.

Inequality 5: _____   Inequality 6: _____

[1 mark]

d. Using inequalities 1 to 6, construct and shade the feasible region for operating the two street stands for one week on the graph on the next page.

[3 marks]

C4. Graphs and relations – extended responses

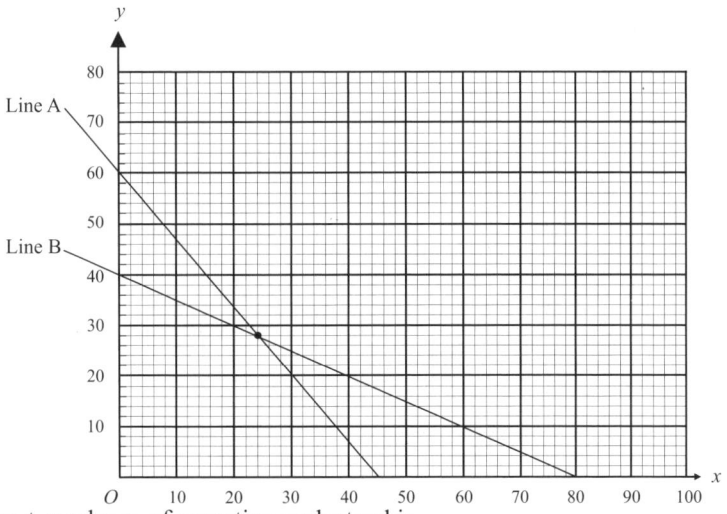

e. The cost, per hour, of operating each stand is
Xenon Street: $100
Yarra Street: $70

Let C be the total cost of operating the two street stands for one week. Write an equation for C in terms of x and y.

[1 mark]

f. Under the conditions described, what is the minimum total cost of operating the street stands for one week?

[2 marks]

Pepi has decided to sell books at each stand in addition to the newspapers and magazines that are being sold.

On average, per hour, each stand will be able to sell
  Xenon Street: 2 books
  Yarra Street: 3 books.

In total, at least 150 books must be sold each week.

g. Write an inequality in terms of x and y for the total number of books that must be sold each week.

Inequality 7: ☐

[1 mark]

h. Draw the line that forms the boundary of the region defined by Inequality 7 on the following graph.

C4. Graphs and relations – extended responses

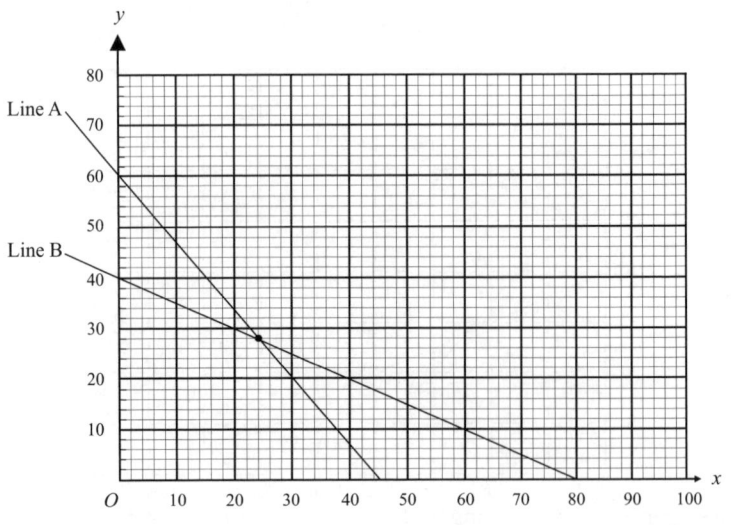

[1 mark]

i. In one particular week Pepi decided to operate the Xenon Street stand for 25 hours and the Yarra Street stand for 32 hours. Will he be able to sell the required number of books if he operates the stands for these hours? Justify your answer.

[1 mark]

j. Pepi is now allowed to operate each street stand for as many hours as he wants each week. He sells newspapers, magazines and books at each stand and meets all sales requirements for these items.
The cost, per hour, of operating each stand increases to
Xenon Street: $120
Yarra Street: $90

    i. Determine the new minimum total cost of operating the street stands for one week.

    ii. Find all solutions for the number of hours per week that each stand can operate so that the minimum total operating cost is achieved.

[1 + 2 = 3 marks]
[VCAA 2005 FM]

Question 323

Harry operates a mobile pet care service. The call-out fee charged depends on the distance he has to travel to tend to a pet. The call-out fees for distances up to 30 km are shown here.

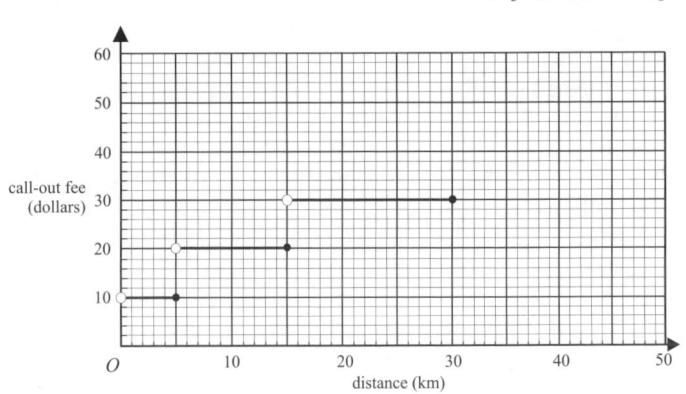

C4. Graphs and relations – extended responses

a. According to this graph
   i. what is the call-out fee to travel a distance of 20 km?
   ii. what is the maximum distance travelled for a call-out fee of $10?

   [1 + 1 = 2 marks]

A call-out fee of $50 is charged to travel distances of more than 30 km but less than or equal to 40 km.

b. **Draw** this information on the graph on the previous page.

   [1 mark]
   [VCAA 2006 FM]

Question 324

In one particular week, Harry began with 50 litres of fuel in the tank of his van. After he had travelled 160 km there were 30 litres of fuel left in the tank of his van. The amount of fuel remaining in the tank of Harry's van followed a linear trend as shown in the graph here.

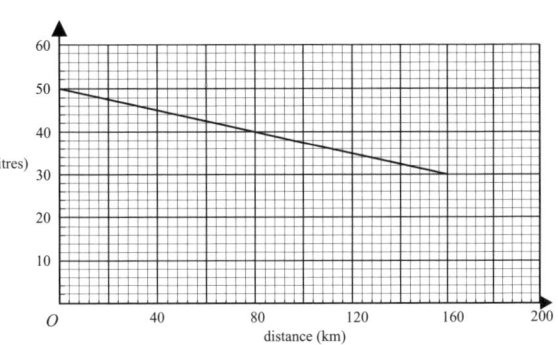

a. Determine the equation of the line shown in the graph above.

   [2 marks]

Assume this linear trend continues and that Harry does not add fuel to the tank of his van.

b. How much **further** will he be able to travel before the tank is empty?

   [1 mark]

Harry stopped to refuel his van when there were 12 litres of fuel left in the tank. He completely filled the tank in $3\frac{1}{2}$ minutes when fuel was flowing from the pump at a rate of 18 litres per minute.

c. How much fuel does the tank hold when it is completely full? Write your answer in litres.

   [1 mark]
   [VCAA 2006 FM]

Question 325

Harry offers dog washing and dog clipping services.
Let x be the number of dogs washed in one day
     y be the number of dogs clipped in one day.
It takes 20 minutes to wash a dog and 25 minutes to clip a dog.
There are 200 minutes available each day to wash and clip dogs.
This information can be written as Inequalities 1 to 3.

C4. Graphs and relations – extended responses

Inequality 1: $x \geq 0$
Inequality 2: $y \geq 0$
Inequality 3: $20x + 25y \leq 200$

a. Draw the line that represents $20x + 25y = 200$ on the graph shown here.

[1 mark]

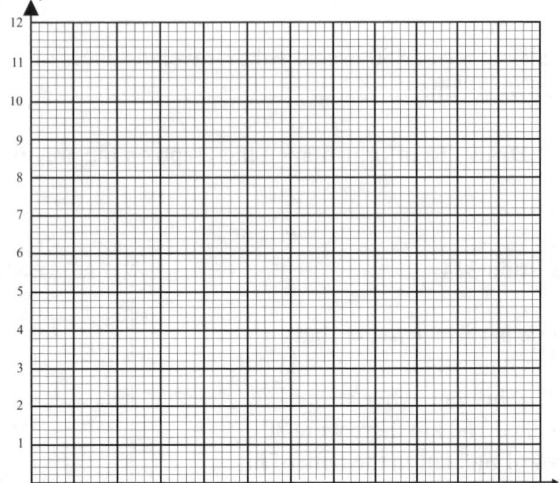

In any one day the number of dogs clipped is at least twice the number of dogs washed.

b. Write an inequality to describe this information in terms of x and y.
Inequality 4: ☐

[1 mark]

c. i. On the graph just given, draw and clearly indicate the **boundaries** of the region represented by Inequalities 1 to 4.
    ii. On a day when exactly five dogs are clipped, what is the maximum number of dogs that could be washed?

[2 + 1 = 3 marks]

The profit from washing one dog is $40 and the profit from clipping one dog is $30. Let P be the total profit obtained in one day from washing and clipping dogs.

d. Write an equation for the total profit, P, in terms of x and y.

[1 mark]

e. i. Determine the number of dogs that should be washed and the number of dogs that should be clipped in one day in order to maximise the total profit.

    ii. What is the maximum total profit that can be obtained from washing and clipping dogs in one day?

[1 + 1 = 2 marks]
[VCAA 2006 FM]

C4. Graphs and relations – extended responses

Question 326

The Goldsmith family are going on a driving holiday in Western Australia.
On the first day, they leave home at 8 am and drive to Watheroo then Geraldton. The distance-time graph shows their journey to Geraldton.

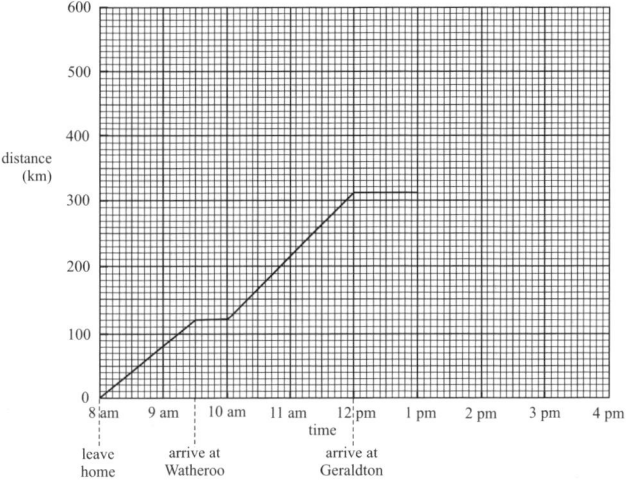

At 9.30 am the Goldsmiths arrive at Watheroo. They stop for a period of time.

a. For how many minutes did they stop at Watheroo?

[1 mark]

After leaving Watheroo, the Goldsmiths continue their journey and arrive in Geraldton at 12pm.

b. What distance (in kilometres) do they travel between Watheroo and Geraldton?

[[1 mark]]

c. Calculate the Goldsmiths' average speed (in km/h) when travelling between Watheroo and Geraldton.

[[1 mark]]

The Goldsmiths leave Geraldton at 1 pm and drive to Hamelin. They travel at a constant speed of 80 km/h for three hours. They do not make any stops.

d. On the graph above, draw a line segment representing their journey from Geraldton to Hamelin.

[1 mark]
[VCAA 2007 FM]

Question 327

The Goldsmiths' car can use either petrol or gas. The following equation models the fuel usage of petrol, P, in litres per 100 km (L/100 km) when the car is travelling at an average speed of s km/h.

C4. Graphs and relations – extended responses

$P = 12 - 0.02s$

The line $P = 12 - 0.02s$ is drawn on the graph here for average speeds up to 110 km/h.

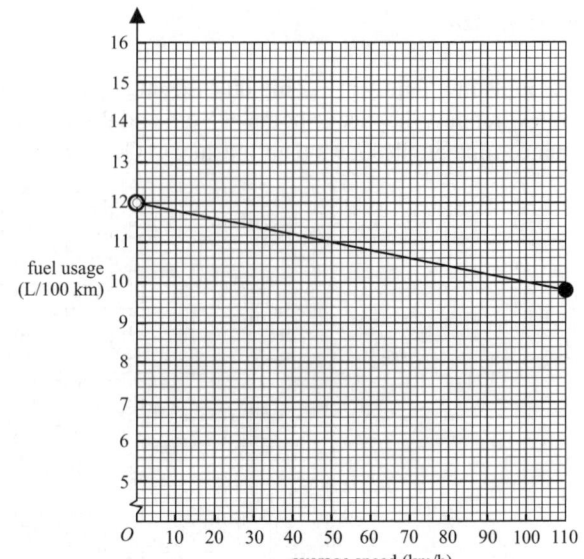

a. Determine how many litres of petrol the car will use to travel 100 km at an average speed of 60 km/h. Write your answer correct to one decimal place.

[1 mark]

The following equation models the fuel usage of gas, G, in litres per 100 km (L/100 km) when the car is travelling at an average speed of s km/h.

$G = 15 - 0.06s$

b. On the axes above, draw the line $G = 15 - 0.06s$ for average speeds up to 110 km/h.

[1 mark]

c. Determine the average speeds for which fuel usage of gas will be less than fuel usage of petrol.

[1 mark]

The Goldsmiths' car travels at an average speed of 85 km/h. It is using gas. Gas costs 80 cents per litre.

d. Determine the cost of the gas used to travel 100 km. Write your answer in dollars and cents.

[2 marks]
[VCAA 2007 FM]

Question 328

Gas is generally cheaper than petrol.
The car must run on petrol for some of the driving time.
Let x be the number of hours driving using gas
  y be the number of hours driving using petrol
The following inequalities, Inequalities 1 to 5, represent the constraints on driving a car over a 24-hour period. Explanations are given for Inequalities 3 and 4.

Inequality 1: $x \geq 0$
Inequality 2: $y \geq 0$

C4. Graphs and relations – extended responses

Inequality 3: $y \leq \frac{1}{2}x$.   The number of hours driving using petrol must not exceed half the number of hours driving using gas.

Inequality 4: $y \geq \frac{1}{3}x$ The number of hours driving using petrol must be at least one third the number of hours driving using gas.

Inequality 5: $x + y \leq 24$

a. Explain the meaning of Inequality 5 in terms of the context of this problem.

[1 mark]

The lines $x + y \leq 24$ and $y = \frac{1}{2}x$ are drawn on the graph shown here.

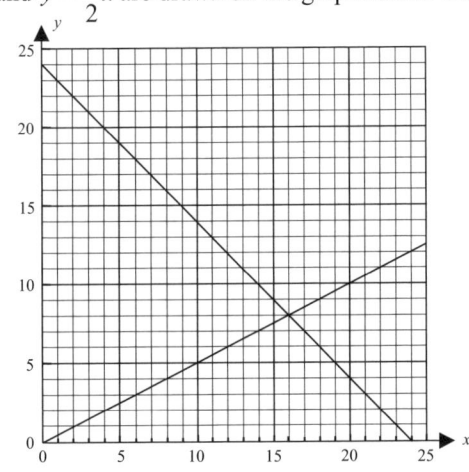

b. On the graph
   i. draw the line $y = \frac{1}{3}x$
   ii. clearly shade the feasible region represented by Inequalities 1 to 5.

[1 + 1 = 2 marks]

On a particular day, the Goldsmiths plan to drive for 15 hours. They will use gas for 10 of these hours.

c. Will the Goldsmiths comply with all constraints? Justify your answer.

[2 marks]

On another day, the Goldsmiths plan to drive for 24 hours. Their car carries enough fuel to drive for 20 hours using gas and 7 hours using petrol.

d. Determine the **maximum** and **minimum** number of hours they can drive using gas while satisfying all constraints.

Maximum = ☐ hours

Minimum = ☐ hours

[2 marks]
[VCAA 2007 FM]

127

C5. Business-related mathematics – extended responses

The following information is for questions 329, 330 and 331.

Rich Aunt Matilda has two nieces, Adele and Briony, and a nephew, Cameron. Aunt Matilda gives each of her nieces and nephew $80 000 to spend as they wish.

Question 329

Adele decides to spend her money as follows.
- $40 000 on a new car
- $40 000 on the latest computer equipment

Adele knows that the car will depreciate by 25% per annum based on the reducing value of the car, whereas the computer equipment will depreciate at a flat rate of $8000 per year.

a. What is the value of the car after
   i. one year?
   ii. three years?

[2 marks]

b. What is the value of the computer equipment after two years?

[1 mark]

c. After how many full years does the depreciated value of the car first exceed the depreciated value of the computer equipment?

[1 mark]

d. Determine the annual percentage flat rate of depreciation applied to the computer equipment.

[1 mark]
[VCAA 2001 FM]

Question 330

Briony decides to purchase a house worth $150 000 by using the $80 000 that she has received, and borrowing $70 000 from a building society. To repay the loan of $70 000, Briony will pay the building society $1500 per month for 10 years.

a. Calculate the total amount of Briony's repayments to the building society.

[1 mark]

b. Determine the total interest on the loan during the 10 years.

[1 mark]

c. Find the annual flat rate of interest charged by the building society. Give your answer correct to one decimal place.

[2 marks]
[VCAA 2001 FM]

C5. Business-related mathematics – extended responses

Question 331

Cameron invests his $80 000 to provide a scholarship valued at $10 000 to the best mathematics student in final year at the school in his town. He does this for as long as the money lasts.

Cameron invests his money at an interest rate of 8.5% per annum compound interest into an annuity. He makes the payments to the winning student each year immediately after the interest is paid into the account. The first scholarship is awarded one year after Cameron first invests his money.

a. How much money is left in the account after the first scholarship award?

[1 mark]

b. Using the annuity formula $A = PR^n - \dfrac{Q(R^n - 1)}{R - 1}$, where $R = 1 + \dfrac{r}{100}$

   i. state the values of Q and R for Cameron's investment

   $Q =$ [   ] $R =$ [   ]

   ii. determine the amount that is left in Cameron's account after 10 years of awarding scholarships. Give your answer to the nearest cent.

[4 marks]

c. What would be the maximum value for each scholarship if they are to be awarded forever?

[1 mark]
[VCAA 2001 FM]

Question 332

Sally's credit union passbook looked like this in June 2002.

Date	Particulars	Deposits	Withdrawals	Balance
01 July 2001	Brought Forward			2400.00
15 Dec 2001	Deposit	1200.00		3600.00
02 Feb 2002	ATM Withdrawal			3000.00
14 May 2002	Interest	85.50		
20 June 2002	ATM Withdrawal		450.00	2635.50

a. Complete this table by filling in

   i. the amount withdrawn on 2 February 2002.

   ii. the account balance for 14 May 2002.

[1 + 1 = 2 marks]

129

C5. Business-related mathematics – extended responses

b. Interest on this account was paid at a rate of 0.3% per month, based on the minimum monthly balance. How much interest did Sally earn for the month of December 2001?

[1 mark]
[VCAA 2002 FM]

Question 333

On 1 July 2002, Sally invested $4000 in a new term deposit that offered a total of $416 interest after two years.

a. What was the annual simple interest rate offered for this term deposit?

[1 mark]

b. An alternative option for Sally had been to invest with a bank at a rate of 4.8% per annum compounding annually. To calculate the total amount in this account after two years with this option, Sally wrote down an equation that looked like this
$$\text{total amount} = 4000 \times c \times c$$
What number should Sally have used for c?

[1 mark]

c. What annual compounding interest rate, correct to two decimal places, would Sally have needed to earn $416 interest in two years on a $4000 investment?

[2 marks]
[VCAA 2002 FM]

Question 334

Sally bought her car five years ago for $23 600 and it is now worth $7000.

Calculate, correct to one decimal place, the percentage annual rate of depreciation of the value of Sally's car over five years

a. on a flat rate basis.

[2 marks]

b. on a reducing balance basis.

[2 marks]
[VCAA 2002 FM]

Question 335

Sally wants to borrow $20 000 for four years. Interest is calculated quarterly on the reducing balance at an annual rate of 8%.

a. Sally decides to use the annuities formula following for calculating this loan.
$$A = PR^n - \frac{Q(R^n - 1)}{R - 1}$$
For Sally's loan

C5. Business-related mathematics – extended responses

    i. $R =$ ☐

    ii. $n =$ ☐

                                                                  [1 + 1 = 2 marks]

b. Sally can afford to repay this loan at $1500 per quarter. Will this enable her to repay the loan in four years? Explain.

                                                                    [2 marks]
                                                                   [VCAA 2002 FM]

Question 336

Brad wants to buy a coffee machine for his cafe. Crazy Bill's normally sells them for $3450, but they have a special discounted price of $3100 for this week.

a. What is the percentage discount? Write your answer correct to one decimal place.

                                                                                       [1 mark]

b. Crazy Bill's offers to sell the machine for the discount price of $3100. The terms of the sale are $200 deposit and $275 per month for 12 months.

    i. What is the total cost of the machine on these terms?

    ii. What is the annual flat rate of interest charged? Write your answer correct to one decimal place.

                                                                     [1 + 2 = 3 marks]

c. Brad sees the same coffee machine for sale at Discount King, also for $3100. The terms of the sale there require no deposit and monthly repayments over two years at an interest rate of 9% per annum, calculated monthly on the reducing balance. The monthly repayments can be determined using the annuities formula:

$$A = PR^n - \frac{Q(R^n - 1)}{R - 1}.$$ The loan is paid out in two years.

    i. What values for n, P and A should be substituted into the annuities formula to determine the monthly repayments?

    ii. What is the monthly repayment for this loan? Write your answer in dollars, correct to two decimal places.

    iii. What is the total cost of the machine from Discount King on these terms? Write your answer correct to the nearest dollar.

                                                                     [2 + 1 + 1 = 4 marks]

d. Whose terms, Crazy Bill's or Discount King's, offer the lowest total cost for the coffee machine? Justify your answer by calculating the difference in total money paid.

                                                                     [1 mark]
                                                                   [VCAA 2003 FM]

C5. Business-related mathematics – extended responses

Question 337

Brad buys the coffee machine with an initial value of $3100. He considers two methods of depreciating the value of the coffee machine.

a. Suppose the value of the machine is depreciated using the reducing balance method over three years and reducing at a rate of 15% per annum. What is the depreciated value of the machine after three years? Write your answer correct to the nearest dollar.

[2 marks]

b. Alternatively, suppose that the machine is depreciated using the unit cost depreciation method. Brad sells 15 000 cups of coffee per year and the unit cost per cup is 3.0 cents. Determine the depreciated value of the machine after three years. Write your answer correct to the nearest dollar.

[2 marks]

c. Brad wants the depreciated value of the machine after three years to be the same when calculated by both methods of depreciation. What would the unit cost per cup have to be for this to occur? Write your answer in cents, correct to one decimal place.

[2 marks]
[VCAA 2003 FM]

Question 338

Remy borrows $650 to buy a digital camera. He fully repays this loan with six monthly repayments of $120.

a. Determine

   i. how much it costs Remy to pay off the loan.
   ii. the total amount of interest Remy pays.

[1 + 1 = 2 marks]

b. For this loan

   i. determine the annual simple interest rate. Write your answer correct to one decimal place.

   ii. the effective interest rate is 37% per annum. Explain why the effective interest rate is greater than the simple interest rate.

[1 + 1 = 2 marks]

c. The $650 price of the camera includes 10% GST (Goods and Services Tax). Determine the price of the camera before the GST was applied. Write your answer correct to the nearest cent.

[1 mark]

C5. Business-related mathematics – extended responses

d. Remy uses his camera for work and he wants to depreciate its value over five years. He can either use a flat rate method of depreciation at the rate of 12% per annum or a reducing balance method of depreciation at the rate of 15% per annum. Which method gives the greater total depreciation over five years? Explain your answer.

[3 marks]
[VCAA 2004 FM]

Question 339

Anna borrows $12 000 at 7.5% interest, per annum, compounding monthly. The loan is to be fully repaid over four years by equal monthly repayments. The monthly repayments can be determined with the annuities formula

$$A = PR^n - \frac{Q(R^n - 1)}{R - 1}$$

a. What values of A, n and P should be substituted into the annuities formula to determine the monthly repayments?

$A = $ ☐

$n = $ ☐

$P = $ ☐

[2 marks]

b. Determine the monthly repayment for this loan. Write your answer correct to the nearest cent.

[1 mark]

c. Determine the total amount of interest paid on the loan after four years.

[1 mark]

d. After six equal repayments have been made, how much has Anna paid off the loan? Write your answer correct to the nearest dollar.

[1 mark]

e. At the end of six months the interest rate increases to 8.0% per annum. Anna still has to completely pay out the balance of the loan within the original period of the loan.

   i. Determine the values of n and P for the remaining period of the loan.

   $n = $ ☐

   $P = $ ☐

   ii. Determine the new monthly repayments that now apply. Write your answer correct to the nearest dollar.

[1 + 1 = 2 marks]
[VCAA 2004 FM]

C5. Business-related mathematics – extended responses

Question 340

Stan bought a $4000 computer under a hire-purchase agreement. He paid $500 deposit and will repay the balance in equal monthly instalments over two years. A flat rate of interest is charged. The total amount Stan will pay for the computer (including the deposit) is $4560.

a. Determine the total amount of interest Stan pays.

[1 mark]

b. Show that the flat rate of interest for this agreement is 8% per annum.

[1 mark]

c. Determine the **effective** rate of interest per annum. Write your answer correct to one decimal place.

[1 mark]

d. Explain why an effective interest rate differs from a flat interest rate.

[1 mark]
[VCAA 2005 FM]

Question 341

For taxation purposes, Stan will depreciate his $4000 computer over five years. At the end of five years the book value of his computer will be $1000.

a. If Stan uses **flat rate depreciation**, determine the annual depreciation rate.

[2 marks]

b. If Stan uses **reducing balance depreciation**, determine the annual depreciation rate. Write your answer correct to one decimal place.

[2 marks]

Question 342

Stan's friend Lena has some money that she wishes to invest for a period of five years. She is considering three investment options.

a. **Investment Option A**
   $10 000 is deposited into an account with an interest rate of 4.8% per annum compounding monthly for five years.

   i. To determine the value of this investment at the end of five years, Lena correctly applies the compound interest formula $A = PR^n$. Write down the values of P, R and n that Lena uses.

   $P =$ [ ] $R =$ [ ] $n =$ [ ]

C5. Business-related mathematics – extended responses

    ii. Calculate the value of Investment Option A at the end of five years. Write your answer correct to the nearest cent.

                                                                                                   [2 + 1 = 3 marks]

b. **Investment Option B**
$4 000 is deposited into an account with an interest rate of 4.8% per annum compounding monthly. At the end of each month, for a period of five years, a further $100 is deposited after interest has been paid. Determine the value of Investment Option B at the end of five years (immediately after the $100 has been deposited). Write your answer correct to the nearest cent.

c. **Investment Option C**
Investment Option B is followed for two years. After this, the amount deposited at the end of each month changes. With the new monthly deposit, Investment Option C is worth $13 000 at the end of the five years.

    i. Find the new amount deposited at the end of each month for the remaining three years. Write your answer correct to the nearest cent.

    ii. Determine the total amount of interest earned by Investment Option C over the five-year period. Write your answer correct to the nearest cent.

                                                                                                  [1 + 2 = 3 marks]
                                                                                                  [VCAA 2005 FM]

Question 343

A company purchased a machine for $60 000. For taxation purposes the machine is depreciated over time. Two methods of depreciation are considered.

a. **Flat rate depreciation**
The machine is depreciated at a flat rate of 10% of the purchase price each year.

    i. By how many dollars will the machine depreciate annually?

    ii. Calculate the value of the machine after three years.

    iii. After how many years will the machine be $12 000 in value?

                                                                                 [1 + 1 + 1 = 3 marks]

b. **Reducing balance depreciation**
The value, V, of the machine after n years is given by the formula
$V = 60\,000 \times (0.85)^n$

    i. By what percentage will the machine depreciate annually?

    ii. Calculate the value of the machine after three years.

C5. Business-related mathematics – extended responses

    iii. At the end of which year will the machine's value first fall below $12 000?

[1 + 1 + 1 = 3 marks]

c. At the end of which year will the value of the machine **first** be less using flat rate depreciation than it be using reducing balance depreciation?

[2 marks]
[VCAA 2006 FM]

Question 344

It is estimated that inflation will average 2% per annum over the next eight years.
If a new machine costs $60 000 now, calculate the cost of a similar new machine in eight years time, adjusted for inflation. Assume no other cost change. Write your answer correct to the nearest dollar.

[1 mark]
[VCAA 2006 FM]

Question 345

The company prepares for this expenditure by establishing three different investments.

a. $7 000 is invested at a simple interest rate of 6.25% per annum for eight years. Determine the total value of this investment at the end of eight years.

[2 marks]

b. $10 000 is invested at an interest rate of 6% per annum compounding quarterly for eight years. Determine the total value of this investment at the end of eight years. Write your answer correct to the nearest dollar.

[1 mark]

c. $500 is deposited into an account with an interest rate of 6.5% per annum compounding monthly. Deposits of $200 are made to this account on the last day of each month after interest has been paid. Determine the total value of this investment at the end of eight years. Write your answer correct to the nearest dollar.

[2 marks]
[VCAA 2006 FM]

Question 346

The company anticipates that it will need to borrow $20 000 to pay for the new machine. It expects to take out a reducing balance loan with interest calculated monthly at a rate of 10% per annum.
The loan will be fully repaid with 24 equal monthly instalments. Determine the total amount of interest that will be paid on this loan. Write your answer to the nearest dollar.

[2 marks]
[VCAA 2006 FM]

C5. Business-related mathematics – extended responses

Question 347

Khan wants to buy some office furniture that is valued at $7000.

a.   i. A store requires 25% deposit. Calculate the deposit.

    The balance is to be paid in 24 equal monthly instalments. No interest is charged.

  ii. Determine the amount of each instalment. Write your answer in dollars and cents.

[1+1 = 2 marks]

b. Another store offers the same $7000 office furniture for $500 deposit and 36 monthly instalments of $220.
  i. Determine the total amount paid for the furniture at this store.

  ii. Calculate the **annual flat rate** of interest charged by this store. Write your answer as a percentage correct to one decimal place.

[1 + 2 = 3 marks]

A third store has the office furniture marked at $7000 but will give 15% discount if payment is made in cash at the time of sale.

c. Calculate the cash price paid for the furniture after the discount is applied.

[1 mark]
[VCAA 2007 FM]

Question 348

Khan decides to extend his home office and borrows $30 000 for building costs. Interest is charged on the loan at a rate of 9% per annum compounding monthly.
Assume Khan will pay only the interest on the loan at the end of each month.

a. Calculate the amount of interest he will pay each month.

[1 mark]

Suppose the interest rate remains at 9% per annum compounding monthly and Khan pays $400 each month for five years.

b. Determine the amount of the loan that is outstanding at the end of five years.
   Write your answer correct to the nearest dollar.

[1 mark]

Khan decides to repay the $30 000 loan fully in equal monthly instalments over five years. The interest rate is 9% per annum compounding monthly.

c. Determine the amount of each monthly instalment. Write your answer correct to the nearest cent.

[1 mark]
[VCAA 2007 FM]

C5. Business-related mathematics – extended responses

Question 349

Khan paid $900 for a fax machine.
This price includes 10% GST (goods and services tax).

a. Determine the price of the fax machine **before** GST was added. Write your answer correct to the nearest cent.

[1 mark]

b. Khan will depreciate his $900 fax machine for taxation purposes. He considers two methods of depreciation.

Flat rate depreciation
Under flat rate depreciation the fax machine will be valued at $300 after five years.
i. Calculate the annual depreciation in dollars.

Unit cost depreciation
Suppose Khan sends 250 faxes a year. The $900 fax machine is depreciated by 46 cents for each fax it sends.
ii. Determine the value of the fax machine after five years.

[1 + 1 = 2 marks]
[VCAA 2007 FM]

Question 350

The books in Khan's office are valued at $10 000.

a. Calculate the value of these books after five years if they are depreciated by 12% per annum using the reducing balance, method. Write your answer correct to the nearest dollar.

[1 mark]]

Khan believes his books should be valued at $4000 after five years.

b. Determine the annual reducing balance depreciation rate that will produce this value. Write your answer as a percentage correct to one decimal place.

[2 marks]
[VCAA 2007 FM]

C6. Networks and decision mathematics – extended responses

Question 351

The network diagram shows the distances, in kilometres, along a series of roads that connect a quarry, Q, with worksites shown as nodes.

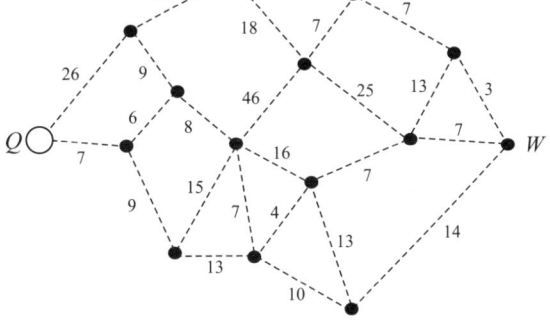

a. One of these worksites is labelled as W.

  i. On the diagram above, clearly draw in the shortest path from the quarry to W.

  ii. Determine the length, in kilometres, of the shortest path between the quarry Q and the worksite W.

  [1 + 1 = 2 marks]

b. The engineer at the quarry wants to visit all worksites in the network. Beginning at Q, he wants to pass through each worksite only once before returning to the quarry.

  i. What mathematical term describes the route the engineer wants to take?

  ii. On the diagram here, clearly draw in a complete route that the engineer could take to visit each worksite only once before returning to the quarry.

  [1 + 2 = 3 marks]

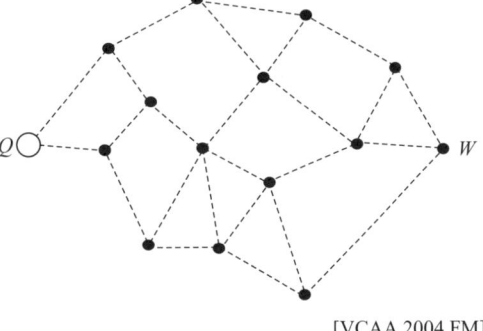

[VCAA 2004 FM]

Question 352

All the activities and their durations (in hours) in a project at the quarry are shown in the network diagram below. The **least time** required for completing this entire project is 30 hours.

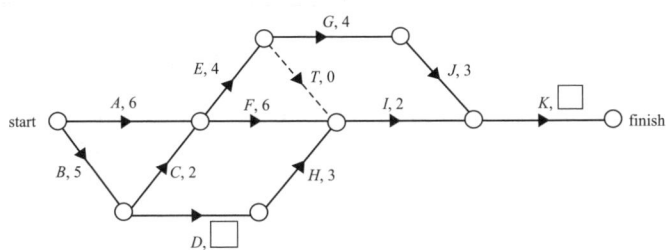

C6. Networks and decision mathematics – extended responses

For each activity in this project, the table below shows the Completion time, the Earliest starting time and the Latest starting time.

a. Complete the missing times in the table.

Activity	Completion time (hours)	Earliest starting time (hours)	Latest starting time (hours)
A	6	0	
B	5	0	0
C	2	5	5
D		5	9
E	4	7	7
F	6	7	
G	4	11	11
H	3	9	13
I	2	13	16
J	3	15	15
K		18	18

b. Write down the critical path for this project.

[1 mark]

To speed up the project, several activities can be dropped from the project. The diagram below shows the activities that must remain in this modified version of the project and their usual completion times.

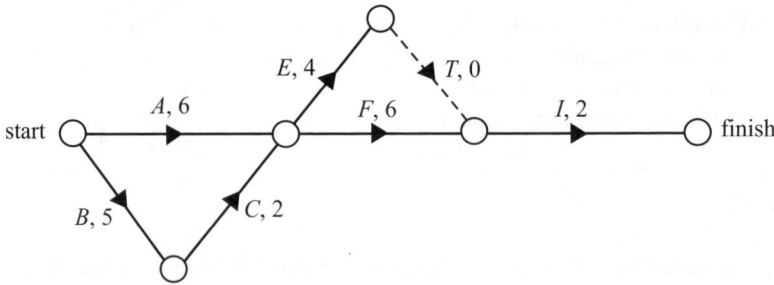

c. Determine the shortest time in which this modified project can be completed.

[1 mark]

The completion time of some of the remaining activities in the **modified project** can be reduced at a cost. The next table shows the reduced times (least possible time to complete an activity after maximum reduction of time). The cost of this reduction, per hour, is also shown.

C6. Networks and decision mathematics – extended responses

Activity	Completion time (hours)	Earliest starting time (hours)	Latest starting time (hours)
A	6	3	50
B	5	4	100
C	2	2	–
E	4	2	20
F	6	4	50
I	2	2	–

d. For this modified project, determine

  i. the activities that should be reduced in time to minimise the completion time of the project.
  ii. the maximum time, in hours, that can be saved by this reduction.
  iii. the minimum cost to achieve this time saving.

[2 + 1 + 1 = 4 marks]
[VCAA 2004 FM]

Question 353

The network diagram here shows the location of a warehouse, W. This warehouse supplies equipment to six factories A, B, C, D, E and F.

The numbers on the edges indicate the shortest distance (in kilometres) to drive along each of the connecting roads.

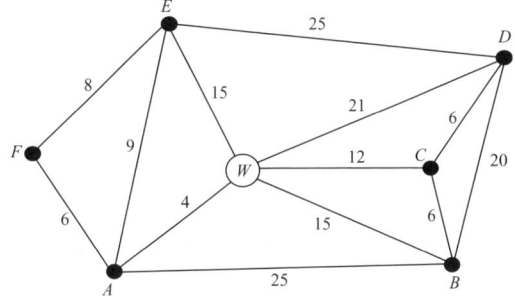

a. The degree of vertex W is ☐

[1 mark]

b. A delivery van is at factory B. It must first make a delivery to factory D and then drive to the warehouse, W. Determine the minimum distance travelled on this journey.

c. A salesman plans to leave factory E, first visit the warehouse, W, and then visit every other factory. He is to visit each location only once. He will will not return to factory E.

  i. Write down the mathematical term used to describe the route that he plans to take.

C6. Networks and decision mathematics – extended responses

ii. On the network diagram shown here, mark in a complete route for his planned journey.

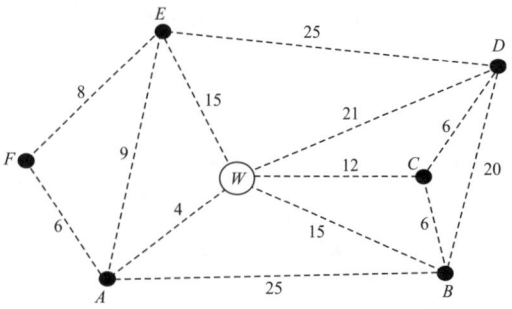

[1 + 1 = 2 marks]

d. The company plans to build an office along one of the roads in the network.

The manager wishes to drive along a route through the network which follows an Euler **circuit**. She will start at the warehouse, *W*.

i. Explain why the journey that the manager plans to take is **not** possible for this network.

ii. A journey that follows an Euler **path**, starting at the warehouse, *W*, is possible for this network. At which vertex will this Euler path end?

[1 + 1 = 2 marks]
[VCAA 2005 FM]

Question 354

One of the delivery vans has to be repaired. This repair usually involves every activity shown in the network diagram below. The duration of each activity, in minutes, is also shown.

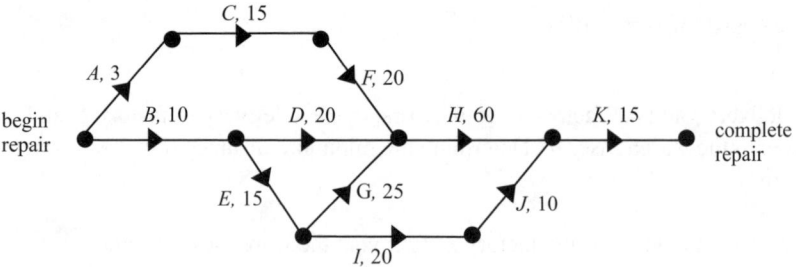

The incomplete table on the next page shows this same information and includes predecessor activities and the earliest starting times (EST).

C6. Networks and decision mathematics – extended responses

Activity	Predecessor(s)	Duration of activity (minutes)	Earliets starting time EST(minutes)
A	–	3	0
C	–	10	0
C	A	15	3
D	B	20	10
E	B	15	10
F	C	20	18
G	E	25	25
H		60	
I	E	20	25
J	I	10	45
K	H, J	15	110

a. Complete the two shaded cells for activity H in the table above.

[2 marks]

b. All activities are required for this repair. What is the minimum time needed to complete this repair?

[1 mark]

c. During the repair to this delivery van it is found that activity F will take longer than the usual 20 minutes.

   i. Explain why the duration of activity F can be increased without delaying this repair.

   ii. What is the maximum duration of activity F that will not delay completion of this repair?

[1 + 1 = 2 marks]

d. Unfortunately further complications arise with this repair. Activity F will now take 40 minutes and activity J will now take 65 minutes. The network diagram below shows the increased duration of these activities.

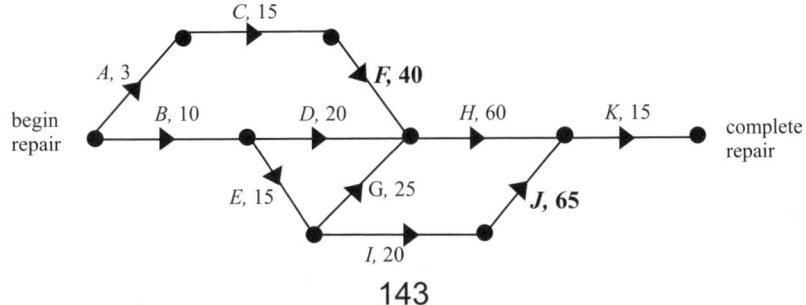

143

C6. Networks and decision mathematics – extended responses

The mechanic also finds that activity *J* now cannot start until activity *G* has been completed.

    i. **Alter the network diagram on the previous page** to allow for this requirement.

    ii. What is the latest starting time (LST) for activity *H* that will not further delay completion of this repair?

    iii. From beginning to completion, what is the minimum time needed to repair this delivery van?

Several activities for this repair can be delayed without increasing the minimum completion time.

    iv. Which of these activities can be delayed for the longest time?

[1 + 1 + 1 + 1 = 4 marks]
[VCAA 2005 FM]

Question 355

George, Harriet, Ian, Josie and Keith are a group of five musicians. They are forming a band where each musician will fill one position only. The following bipartite graph illustrates the positions that each is able to fill.

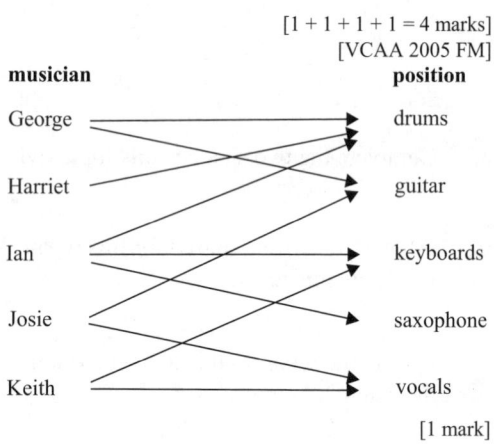

a. Which musician must play the guitar?

[1 mark]

b. Complete the table showing the positions that the following musicians **must** fill in the band.

Person	Position
Harriet	
Ian	
Keith	

[2 marks]
[VCAA 2006 FM]

C6. Networks and decision mathematics – extended responses

Question 356

The five musicians compete in a music trivia game. Each musician competes once against every other musician. In each game there is a winner and a loser.
The results are represented in the dominance matrix, Matrix 1, and also in the **incomplete** directed graph below. On the directed graph an arrow from Harriet to George shows that Harriet won against George.

Matrix 1
$$\begin{array}{c} & \begin{array}{ccccc} G & H & I & J & K \end{array} \\ \begin{array}{c} G \\ H \\ I \\ J \\ K \end{array} & \left[\begin{array}{ccccc} \mathbf{0} & 0 & 1 & 0 & 1 \\ 1 & \mathbf{0} & 1 & 1 & 0 \\ 0 & 0 & \mathbf{0} & 1 & 0 \\ 1 & 0 & 0 & \mathbf{0} & 0 \\ 9 & 1 & 1 & 1 & \mathbf{0} \end{array} \right] \end{array}$$

Directed graph

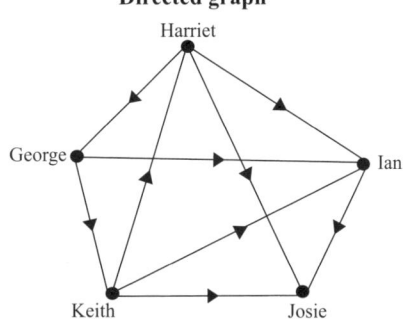

a. Explain why the figures in bold in Matrix 1 are all zero.
[1 mark]

One of the edges on the directed graph is missing.

b. Using the information in Matrix 1, **draw** in the missing edge on the directed graph above and clearly show its **direction**.
[1 mark]

The results of each trivia contest (one-step dominances) are summarised as follows.

Musician	Dominance value (wins)
George	2
Harriet	3
Ian	1
Josie	1
Keith	3

In order to rank the musicians from first to last in the trivia contest, two-step (two-edge) dominances will be considered. The following **incomplete** matrix, Matrix 2, shows two-step dominances.

C6. Networks and decision mathematics – extended responses

Matrix 2

$$\begin{array}{c} \\ G \\ H \\ I \\ J \\ K \end{array} \begin{bmatrix} G & H & I & J & K \\ 0 & 1 & 1 & 2 & 0 \\ 1 & 0 & 1 & 1 & 1 \\ 1 & 0 & 0 & 0 & 0 \\ 0 & 0 & 1 & 0 & 1 \\ 2 & 0 & 1 & x & 0 \end{bmatrix}$$

c. Explain the two-step dominance that George has over Ian.

[1 mark]

d. Determine the value of the entry x in Matrix 2.

[1 mark]

e. Taking into consideration both the one-step and two-step dominances, determine which musician was ranked first and which was ranked last in the trivia contest.

[2 marks]
[VCAA 2006 FM]

Question 357

The five musicians are to record an album. This will involve nine activities. The activities and their immediate predecessors are shown in the following table. The duration of each activity is not yet known.

Activity	Immediate predecessors
A	–
B	–
C	–
D	A
E	B
F	B
G	D, E
H	F
I	G, H

C6. Networks and decision mathematics – extended responses

a. Use the information in the table above to complete the network below by including activities *G*, *H* and *I*.

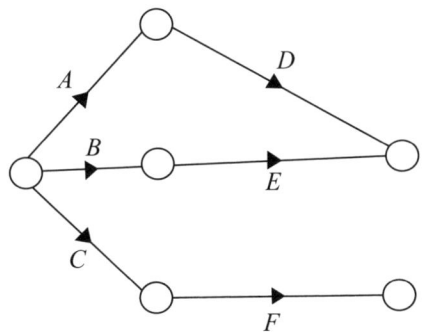

[2 marks]

There is only one critical path for this project.

b. How many **non-critical** activities are there?

[1 mark]

The following table gives the earliest start times (EST) and latest start times (LST) for **three of the activities only**. All times are in hours.

Activity	EST	LST
A	0	2
C	0	1
I	12	12

c. Write down the critical path for this project.

[1 mark]

The minimum time required for this project to be completed is 19 hours.

d. What is the duration of activity *I*?

[1 mark]

The duration of activity *C* is 3 hours.

e. Determine the maximum combined duration of activities *F* and *H*.

[1 mark]
[VCAA 2006 FM]

Question 358

A new housing estate is being developed. There are five houses under construction in one location. These houses are numbered as points 1 to 5.

C6. Networks and decision mathematics – extended responses

The builders require the five houses to be connected by electrical cables to enable the workers to have a supply of power on each site.

a. What is the **minimum** number of edges needed to connect the five houses?

[1 mark]

b. On the diagram on the previous page, **draw** a connected graph with this number of edges.

[[1 mark]]
[VCAA 2007 FM]

Question 359

The estate has large open parklands that contain seven large trees.
The trees are denoted as vertices A to G on the network diagram. Walking paths link the trees as shown.

The numbers on the edges represent the lengths of the paths in metres.

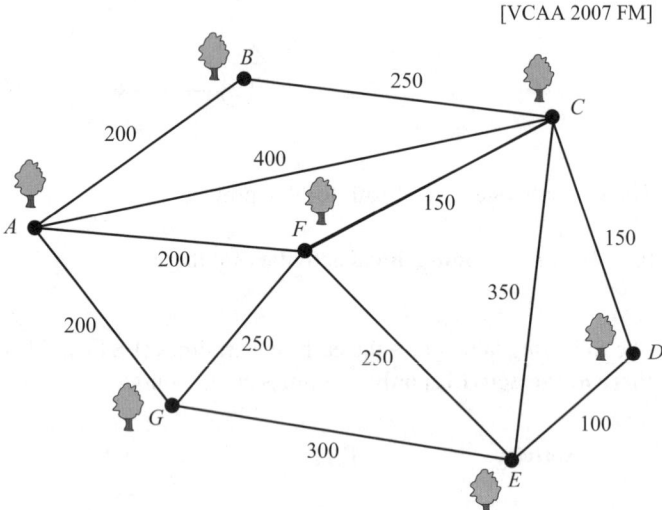

a. Determine the sum of the degrees of the vertices of this network.

[[1 mark]]

b. One day Jamie decides to go for a walk that will take him along each of the paths between the trees. He wishes to walk the minimum possible distance.

   i. State a vertex at which Jamie could begin his walk.

   ii. Determine the total distance, in metres, that Jamie will walk.

[1+1=2 marks]

Michelle is currently at F. She wishes to follow a route that can be described as the **shortest** Hamiltonian circuit.

c. Write down a route that Michelle can take.

[1 mark]
[VCAA 2007 FM]

Question 360

As an attraction for young children, a miniature railway runs throughout the new housing estate. The trains travel through stations that are represented by nodes on the following directed network diagram. The number of seats available for children, between each pair of stations, is indicated beside the corresponding edge.

148

C6. Networks and decision mathematics – extended responses

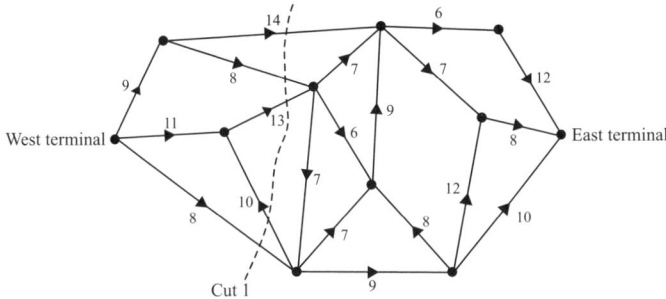

Cut 1, through the network, is shown in the diagram above.

a. Determine the capacity of Cut 1.

[1 mark]

b. Determine the maximum number of seats available for children for a journey that begins at the West Terminal and ends at the East Terminal.

[1 mark]

On one particular train, 10 children set out from the West Terminal.
No new passengers board the train on the journey to the East Terminal.

c. Determine the maximum number of children who can arrive at the East Terminal on this train.

[1 mark]
[VCAA 2007 FM]

Question 361

A community centre is to be built on the new housing estate. Nine activities have been identified for this building project. The directed network here shows the activities and their completion times in weeks.

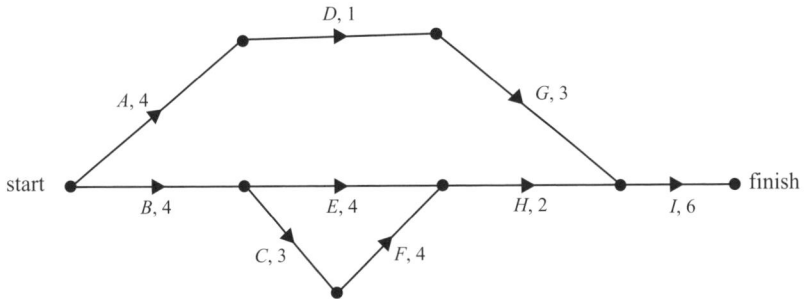

a. Determine the minimum time, in weeks, to complete this project.

[1 mark]

b. Determine the slack time, in weeks, for activity D.

[2 marks]

C6. Networks and decision mathematics – extended responses

The builders of the community centre are able to speed up the project. Some of the activities can be reduced in time at an additional cost. The activities that can be reduced in time are A, C, E, F and G.

c. Which of these activities, if reduced in time individually, would **not** result in an earlier completion of the project?

[1 mark]

The owner of the estate is prepared to pay the additional cost to achieve early completion. The cost of reducing the time of each activity is $5000 per week.
The maximum reduction in time for each one of the five activities, A, C, E, F, G, is 2 weeks.

d. Determine the minimum time, in weeks, for the project to be completed now that certain activities can be reduced in time.

[1 mark]

e. Determine the minimum additional cost of completing the project in this reduced time.

[1 mark]
[VCAA 2007 FM]

C7. Matrices – extended responses

Question 362

A manufacturer sells three products, A, B and C, through outlets at two shopping centres, Eastown (E) and Noxiand (N). The number of units of each product sold per month through each shop is given by the matrix Q, where

$$Q = \begin{bmatrix} A & B & C \\ 2500 & 3400 & 1890 \\ 1765 & 4588 & 2456 \end{bmatrix} \begin{matrix} \\ E \\ N \end{matrix}$$

a. Write down the order of matrix Q.
[1 mark]

The matrix P, shown below, gives the selling price, in dollars, of products A, B, C.

$$P = \begin{bmatrix} 14.50 \\ 21.60 \\ 19.20 \end{bmatrix} \begin{matrix} A \\ B \\ C \end{matrix}$$

b. **i.** Evaluate the matrix M, where $M = QP$.

ii. What information does the elements of matrix M provide?
[1 + 1 = 2 marks]

c. Explain why the matrix PQ is **not** defined.
[1 mark]
[VCAA 2006 FM]

Question 363

A new shopping centre called Shopper Heaven (S) is about to open. It will compete for customers with Eastown (E) and Noxland (N). Market research suggests that each shopping centre will have a regular customer base but attract and lose customers on a weekly basis as follows.

80% of Shopper Heaven customers will return to Shopper Heaven next week
12% of Shopper Heaven customers will shop at Eastown next week
8% of Shopper Heaven customers will shop at Noxland next week

76% of Eastown customers will return to Eastown next week
9% of Eastown customers will shop at Shopper Heaven next week
15% of Eastown customers will shop at Noxland next week
85% of Noxland customers will return to Noxland next week
10% of Noxland customers will shop at Shopper Heaven next week
5% of Noxland customers will shop at Eastown next week

C7. Matrices – extended responses

a. Enter this information into transition matrix T as indicated below (express percentages as proportions, for example write 76% as 0.76).

$$T = \begin{bmatrix} & & \\ & & \\ & & \end{bmatrix} \begin{matrix} S \\ E \\ N \end{matrix} \text{ next week}$$

$$\begin{matrix} \text{this week} \\ S \quad E \quad N \end{matrix}$$

[2 marks]

During the week that Shopper Heaven opened, it had 300 000 customers. In the same week, Eastown had 120 000 customers and Noxland had 180 000 customers.

b. Write this information in the form of a column matrix, K_0, as indicated below.

$$K_0 = \begin{bmatrix} \\ \\ \end{bmatrix} \begin{matrix} S \\ E \\ N \end{matrix}$$

c. Use T and K_0 to write and evaluate a matrix product that determines the number of customers expected at each of the shopping centres during the following week.

[2 marks]

d. Show by calculating at least two appropriate state matrices that, in the long term, the number of customers expected at each centre each week is given by the matrix

$$K = \begin{bmatrix} 194\,983 \\ 150\,513 \\ 254\,504 \end{bmatrix}$$

[2 marks]
[VCAA 2006 FM]

Question 364

Market researchers claim that the ideal number of bookshops (x), sports shoe shops (y) and music stores (z) for a shopping centre can be determined by solving the equations

$$2x + y + z = 12$$
$$x - y + z = 1$$
$$2y - z = 6$$

a. Write the equations in matrix form using the following template.

$$\begin{bmatrix} & & \\ & & \\ & & \end{bmatrix} \begin{bmatrix} \\ \\ \end{bmatrix} = \begin{bmatrix} \\ \\ \end{bmatrix}$$

[1 mark]

C7. Matrices – extended responses

b. Do the equations have a unique solution? Provide an explanation to justify your response.

[1 mark]

c. Write down an inverse matrix that can be used to solve these equations.

[1 mark]

d. Solve the equations and hence write down the estimated ideal number of bookshops, sports shoe shops and music stores for a shopping centre.

[1 mark]
[VCAA 2006 FM]

Question 365

The table below displays the energy content and amounts of fat, carbohydrate and protein contained in a serve of four foods: bread, margarine, peanut butter and honey.

Food	Energy content (kilojoules/serve)	Fat (grams/serve)	Carbohydrate (grams/serve)	Protein (grams/serve)
Bread	531	1.2	20.1	4.2
Margarine	41	6.7	0.4	0.6
Peanut butter	534	10.7	3.5	4.6
Honey	212	0	12.5	0.1

a. Write down a 2×3 matrix that displays the fat, carbohydrate and protein content (in columns) of bread and margarine.

[1 mark]

b. A and B are two matrices defined as follows.

$$A = [2 \ 2 \ 1 \ 1] \qquad B = \begin{bmatrix} 531 \\ 41 \\ 534 \\ 212 \end{bmatrix}$$

   i. Evaluate the matrix product AB.
   ii. Determine the order of matrix product BA.

Matrix A displays the number of servings of the four foods: bread, margarine, peanut butter and honey, needed to make a peanut butter and honey sandwich.
Matrix B displays the energy content per serving of the four foods: bread, margarine, peanut butter and honey.

   iii. Explain the information that the matrix product AB provides.

[1+1+1=3 marks]

C7. Matrices – extended responses

c. The number of serves of bread (*b*), margarine (*m*), peanut butter (*p*) and honey (*h*) that contain, in total, 53 grams of fat, 101.5 grams of carbohydrate, 28.5 grams of protein and 3568 kilojoules of energy can be determined by solving the matrix equation.

$$\begin{bmatrix} 1.2 & 6.7 & 10.7 & 0 \\ 20.1 & 0.4 & 3.5 & 12.5 \\ 4.2 & 0.6 & 4.6 & 0.1 \\ 531 & 41 & 534 & 212 \end{bmatrix} \begin{bmatrix} b \\ m \\ p \\ h \end{bmatrix} = \begin{bmatrix} 53 \\ 101.5 \\ 28.5 \\ 3568 \end{bmatrix}$$

Solve the matrix equation to find the values *b*, *m*, *p* and *h*.

[2 marks]
[VCAA 2007 FM]

Question 366

To study the life-and-death cycle of an insect population, a number of insect eggs (*E*), juvenile insects (*J*) and adult insects (*A*) are placed in a closed environment. The initial state of this population can be described by the column matrix

$$S_0 = \begin{bmatrix} 400 \\ 200 \\ 100 \\ 0 \end{bmatrix} \begin{matrix} E \\ J \\ A \\ D \end{matrix}$$

A row has been included in the state matrix to allow for insects and eggs that die (*D*).

a. What is the total number of insects in the population (including eggs) at the beginning of the study?

[1 mark]

In this population
- eggs may die, or they may live and grow into juveniles
- juveniles may die, or they may live and grow into adults
- adults will live a period of time but they will eventually die.

In this population, the adult insects have been sterilised so that no new eggs are produced. In these circumstances, the life-and-death cycle of the insects can be modelled by the transition matrix

$$\begin{matrix} & \text{this week} & \\ & E \quad J \quad A \quad D & \end{matrix}$$
$$\begin{bmatrix} 0.4 & 0 & 0 & 0 \\ 0.5 & 0.4 & 0 & 0 \\ 0 & 0.5 & 0.8 & 0 \\ 0.1 & 0.1 & 0.2 & 1 \end{bmatrix} \begin{matrix} E \\ J \\ A \\ D \end{matrix} \quad \text{next week}$$

b. What proportion of **eggs turn into juveniles** each week?

C7. Matrices – extended responses

c. i. Evaluate the matrix product $S_1 = TS_0$

$$S_1 = tS_0 = \begin{bmatrix} & & \end{bmatrix} \begin{bmatrix} E \\ J \\ A \\ D \end{bmatrix}$$

  ii. Write down the number of live **juveniles** in the population after one week.

  iii. Determine the number of live **juveniles** in the population after four weeks. Write your answer correct to the nearest whole number.

  iv. After a number of weeks there will be no live eggs (less than one) left in the population. When does this first occur?

  v. Write down the exact steady-state matrix for this population.

$$S_{\text{steady state}} = \begin{bmatrix} & & \end{bmatrix} \begin{bmatrix} E \\ J \\ A \\ D \end{bmatrix}$$

[1+1+1+1+1=5 marks]

d. If the study is repeated with unsterilised adult insects, eggs will be laid and potentially grow into adults. Assuming 30% of adults lay eggs each week, the population matrix after one week, S_1, is now given by

$$S_1 = TS_0 + BS_0$$

where $B = \begin{bmatrix} 0 & 0 & 0.3 & 0 \\ 0 & 0 & 0 & 0 \\ 0 & 0 & 0 & 0 \\ 0 & 0 & 0 & 0 \end{bmatrix}$ and $S_0 = \begin{bmatrix} 400 \\ 200 \\ 100 \\ 0 \end{bmatrix} \begin{matrix} E \\ J \\ A \\ D \end{matrix}$

  i. Determine S_1

$$S_1 = \begin{bmatrix} & & \end{bmatrix} \begin{bmatrix} E \\ J \\ A \\ D \end{bmatrix}$$

[1+1+1+1+1=5 marks]

This pattern continues. The population matrix after n weeks, S_n, is given by

$$S_n = TS_{n-1} + BS_{n-1}$$

  ii. Determine the number of live eggs in this insect population after two weeks.

[1+1=2 marks]
[VCAA 2007 FM]

Solutions: A1

Question 1 D

First sort the data into order: −5.5, −4.7, −4.6, −4.4, −1.1, 0.3, 2.9. Second, count them: there are 7 items. Therefore the median is the fourth number, −4.4%.

Question 2 E

The most negative number is −5.5. The largest number is 2.9. Therefore the range is 2.9 − (−5.5) = 8.4%.

Question 3 D

Both 52 and 76 are 12 units from the mean. This is one standard deviation away from the mean. We expect about 67% of the scores to be within one standard deviation of the mean. Thus, 68% of 2500 is 1700.

Question 4 C

12 people have an average weight of 72 kg. Their total weight is 12 × 72 = 864 kg.

Question 5 C

When you add a number to a list of 12 numbers, the mean increases by that number but the standard deviation will not change. Thus, the mean will be 74 and the standard deviation will be 5.

Question 6 B

The number of standard-sized houses is 14 + 71 + 47 = 132 houses.
The number of them rated as high is 47.
Thus, the percentage is $\frac{47}{132} \times 100 = 35.6\%$.

Question 7 B

Both 'level of water usage' and 'size of house' use words in their description. Therefore they are categorical variables. Note that the numbers do not 'belong' to either of the variables. They are a count of items in a combination of categories.

Question 8 B

Count how many numbers there are on the right-hand side of the ordered stemplot: there are 26.

Question 9 B

The 'middle number' is between the 13th and 14th number (it is the number 36). There are 12 numbers either side. Group them with the middle number so there are an odd number of numbers 'either side of the centre'. The middle of those will be 30 and 43, giving an IQR of approximately 13.

Question 10 C

An egg with weight 95 g is 10 g above the mean. This is 2 standard deviations. Now, 95% of scores are within 2 standard deviations of the mean. Thus, 5% are outside that range. Of those half (2.5%) are above the upper limit.

Question 11 D

'At least' one DVD players means 1, 2, 3, 4 or 5 DVD players. Since the 'opposite case' is zero DVD players, and there are 6 households with zero DVDs, 20 − 6 = 14 must have DVDs. 14 out of 20 is 70%.

Question 12 C

'Multiply across' in this table and add down. You should get

Solutions: A1

```
0 | 6 | 0
1 | 9 | 9
2 | 3 | 6
3 | 1 | 3
4 | 0 | 0
5 | 1 | 5
       23
```

That is the total number of DVDs (23). Now divide by 20, the number of households, and you get 1.15.

Question 13 C

It is rather useful to complete the table. Look at each row and column to ensure they 'all add up'. You have to be a little creative with your calculator as you fill in the table.

44	9	8	61
16	**24**	18	58
10	**41**	47	**98**
70	**74**	73	217

The figures in bold replace the 'gaps'. There are 61 who never use the Internet.

Question 14 E

From the table just given, there are 24 in the 7-10 group who sometimes use the Internet. Since there are 74 in the 7-10 group, this is $\frac{24}{74} \times \frac{100}{1} = 32.43 \approx 32\%$.

Question 15 B

There are 10 cars. The 5th and 6th have speeds of 62.8 and 62.6, making the median speed 62.7 km/h.

Question 16 E

7 of the drivers are over 25 years.

This means $\frac{7}{10} \times \frac{100}{1} = 70\%$.

Question 17 D

Since the longest tail is to the left, then this is negatively skewed. Those small bars to the far right represent outliers. Therefore, answer D.

Question 18 C

First, look at the wording and the table very carefully during reading time. Such matters as the word 'never' should not be missed. In the '2 column', there are 30 households. Of them, 3 never use text messages. This is 10% of the 30.

Question 19 A

Multiply the top line numbers (0, 1, 2, 3) by the bottom line numbers (34, 78, 30 and 12). That gives the total number of phones. Then divide by 154: $\frac{78 + 60 + 36}{154} = 1.13$.

Question 20 C

Expressed on a diagram:

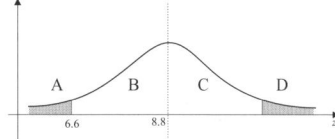

Regions B and C add up to 68% of the total because they are start and finish 1 standard deviation from the mean (8.8 ± 2.2). Thus, A and D total 32% meaning that region A is 16% of the total.

Question 21 D

Moving from the top boxplot to the lower boxplot, the shape of the data becomes narrower (less variable) and more to the right on the graph scale (increases on average).

Solutions: A1

Question 22 D

Temperature can take 'any value' and is a numerical variable. The town can only be Beachside or Flattown and is categorical.

Question 23 B

Minimum temperatures: Beachside has a minimum 38° and maximum of 15°. The range is, therefore, 38 – 15 = 23°.

Question 24 E

If you ignore those first two numbers for Flattown (the 18 and 19) then the numbers are symmetrically centred around the 30s. Those two numbers are outliers. Proving they are outliers (more than 1.5IQR below the mean) is a worthwhile exercise - but not during the exam…

Question 25 C

Note that 48 is 1.5 unites above the mean value of 49.5. That is, one standard deviation (SD) below. Now, it is known that the mean ± one SD contains 68% of the data. As a diagram:

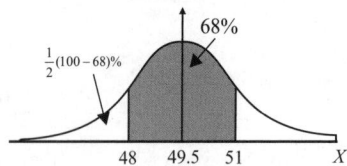

The white area to the left is 16% of the total. 16% of 400 is 64 boys.

Question 26 D

There are 6 columns for those who got 30 or more. Their heights are added:
  7 + 11 + 14 + 16 + 18 + 12
= 78%

Question 27 B

For the median, half the numbers must be above and half below its value. That is, we want the heights of the columns to the left (or to the right) of the median to add up to 50. Here, it is easier to add the right-hand columns:
Last column: 12
Last two columns: 12 + 18 = 30
Last three: 12 + 18 + 16 = 46
Last four: 12 + 18 + 16 + 14 = 60
So the median is somewhere inside the fourth last column. That is, in the region 40 and 45.

Question 28 A

The most 'popular number' along the horizontal axis is 1. There are more one bedroom apartments than any of the others.

Question 29 B

The median is found by considering all of those dots put in a horizontal row. The 11th dot will be the median. Now count from the left. The 11th dot represents a two-bedroom apartment.

Question 30 A

To convert to from x-values to standardised scores (z-values) use the equation
$$z = \frac{x - \mu}{\sigma}$$
$$= \frac{56 - 67}{10.2}$$
$$= -1.0784$$
$$\approx -1.08$$

Question 31 D

The data in the question is transferred to this diagram;

Solutions: A1

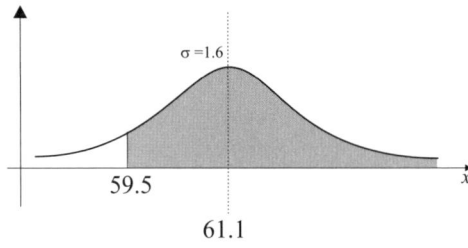

Calculate how far it is from 59.9 to 61.1. It is exactly one standard deviation Now, about 68% of normal data is within one standard deviation of the mean. Therefore, the white area in that diagram is $\frac{1}{2}(100-68) = 16\%$ of the whole area. That means the shaded area is $100 - 16 = 84\%$.

Question 32 C

'A diameter greater than 14 mm' means data values on the upper boxplot starting at Q_3 and going to the right. That is, 25%.

Question 33 C

The lower boxplot is wider and generally lower in corresponding values than the values in the upper boxplot. That leaves just alternatives **B** and **C**. The width of the lower boxplot makes it more variable than the upper boxplot. Alternative **C** seems far more sensible.

Question 34 E

The fact that there are only 3 categories makes it difficult to use a histogram, stemplot or boxplot. Since there are only 9 combinations for any one car, there could only be 9 'dots' on a scatterplot. This is a possibility I guess, but a percentaged, segmented bar chart seems sensible.

Solutions: A2

Question 35 B

One method is to look for 35 minutes underneath the graph axes between 30 and 40. Follow up to the graph and across to the left. I got a value of 4.5 giving B, 4.6, the best chance.
Alternatively, substitute study time = 35 into the equation
∴ errors = $8.8 - 0.120 \times 35$
= 4.6 errors.

Question 36 A

If the coefficient of determination is 0.8198 (look under the graph in the question), then r must be $\sqrt{0.8198}$. That is, 0.9054. Oh, come to think of it, it could be ±. But since the graph is clearly leaning backwards r must be negative… –0.0954, nearest to –0.91.

Question 37 C

One way of looking at this problem is that the 'dots need stretching to the right'. To achieve this, squaring the x-values would be a good idea. End of question.

Being a little curious (and willing to waste large amounts of time), I calculated the correlation coefficient (r) for each of the alternative answers (and one more) to compare…

- a $\frac{1}{x}$ transformation? $r = -0.76$
  (provided you leave out the first point).
- a $\frac{1}{y}$ transformation? $r = -0.63$
- a x^2 transformation? $r = -0.97$
- a log x transformation? $r = -0.87$
  (provided you leave out the first point).
- a log y transformation? $r = -0.81$
- do nothing? $r = -0.93$. Mmm, close…

Question 38 D

Since fitness level (below average, average, above average) is a variable 'with words', it is a categorical variable. So there goes any chance of a line, scattergraph or time series being a good idea.

Since there are three categories, the back-to-back idea becomes impractical. That leaves the good old histogram and the three parallel boxplots idea. The histogram does not work because there is not really the concept of a count for any category. But the parallel boxplot idea would be ideal, as the variation of result within each category can be shown in this way.

Question 39 A

The 'y-intercept' on this graph is about 210. That immediately means the answer is either A or B. The slope is negative, eliminating B. There, that took less than 2 minutes…

Question 40 C

The coefficient of determination is the r-value squared. That gives 0.857476 or, correct to 4 decimal places, 0.8575. (Squaring negative numbers results in a positive number...)

Question 41 B

To fit the three median line to the data, I went through these steps:

1. I divided the data into three groups. You choose the group by 'walking along the horizontal axis and counting the plus signs above your head'. You can see the grouping in this diagram:

Solutions: A2

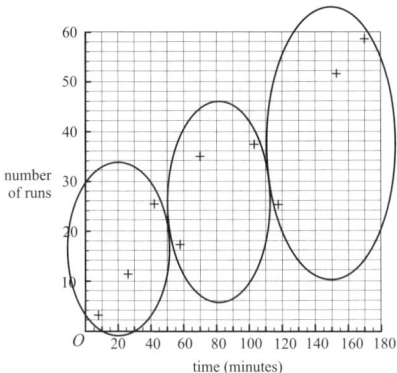

2. Mark the three 'summary' points. That is, the three 'median' points.

3. Join points 1 and 3 as shown below and move the line one-third of the way to summary point 2.

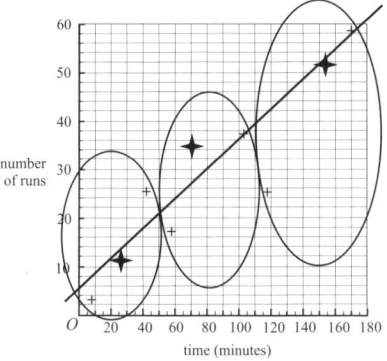

To find the slope of that line, measure its 'rise over run' before you shift it. I measured the gradient as $\frac{57}{180} \approx 0.316$, closest to 0.3. Yes, yes, shifting it was a bit of a waste of time.

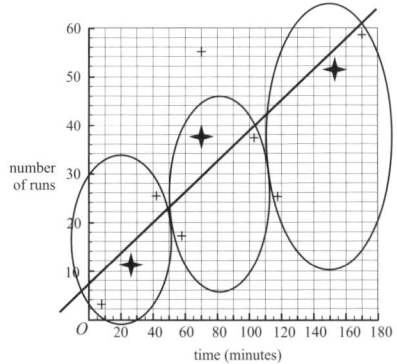

Question 42 E

If (70, 55) replaces (70, 35), then the middle 'summary point' moves up. As the gradient of the three-median regression line is based on the first and third summary points, it will not change because of this correction.

Question 43 C

This question took a little longer than I had hoped. In my 'spreadsheet' environment in my calculator, I typed the Length values into column 1 and the Wingspan numbers into column 2. 'LINREG' gave me an equation for the line plus the bonus of an 'r' value of 0.9681.

Question 44 B

Many will answer this question without the slightest need for the computation. For others however, try this: Put the equation $y = -2.99 + 0.96x$ into the graph equation area of your calculator, set the table to start at $x = 1$ and increase by 1, and then press 2nd graph. The table will show a set of x-numbers increasing by 1 unit at a time and y-values that increase by 0.96 units at a time.

Question 45 A

For the equation $y = a + bx$, we know that
$$b = r\frac{s_y}{s_x}$$
$$= -0.5675 \times \frac{6.98}{2.61}$$
$$= -1.517682$$
$$\approx -1.52$$

Also, $a = \bar{y} - b\bar{x}$
$$= 23.93 - (-1.517682) \times 4.56$$
$$= 30.851086$$
$$\approx 30.9$$

Our equation becomes $y = 30.9 - 1.52x$.

Solutions: A2

Question 46 C

By definition,
    residual = actual − predicted.
Here, a waist measurement of 80 matches a weight (from the table) of 67 kg. Using the regression equation, we get
    weight = −20 + 1.11× waist
= −20 + 1.11× 80
= 68.8
The residual is 67 − 68.8
= −1.8
This is closest to −2 cm.

1	1.08
2	1.40
3	1.52
4	1.76
5	1.99
6	2.23
7	2.54
8	2.60
9	2.94

Question 47 A

This is a graphical calculator question. I put the square of the x-values in the first column. That is, 0, 1, 4, 9, ... 81. Then I put the y-numbers in the second column. That is, the 5, 7, 14, ... 238. The regression is found by pressing LINREG or the equivalent. I got $a = 2.9387$ and $b = 7.1470$, (and wasting just a little time here, $r = 0.996$). Now, remember that we are dealing with x^2 now, not x. This all means $y = 7.1 + 2.9x^2$ is the closest answer.

I applied the least squares routine from my graphical calculator and obtained a gradient of 0.225707. The only equation that has that gradient is alternative **D**. That is, $\log y = 0.88 + 0.23x$.
Note: some may prefer to take any two points, convert y to $\log(y)$ and find the gradient between them. Actually that would be much faster.

Question 48 C

Written as a percentage. the coefficient of determination is 100×0.9034^2
≈   81.61%
≈   81.6%

Question 49 D

The gradient of the graph is 0.87. That means that there is a 0.87 mm increase in *length* for each 1 mm increase in *diameter*.

Question 50 D

The table in the exam question needs to be altered. Each of the y-values is replaced with $\log_{10} y$. The results are listed here − vertically for convenience:

Solutions: A3

Question 51 B

Fitting the 3-median line goes like this:

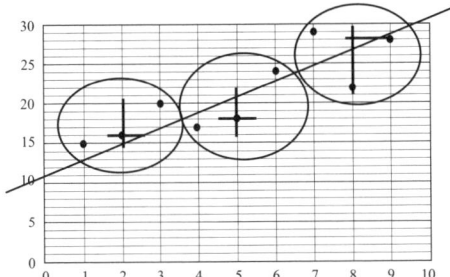

The first and third 'summary points' are used to draw the line which is then shifted by one-third towards the middle summary point. (This is not needed here because it does not change the gradient of the line.) I measured the gradient using the largest triangle.
They are (0, 11) and (9, 29). The gradient between them is:
$$\frac{29-11}{9-0} = \frac{18}{9} = 2$$

Question 52 C

When that dot is moved up like this:

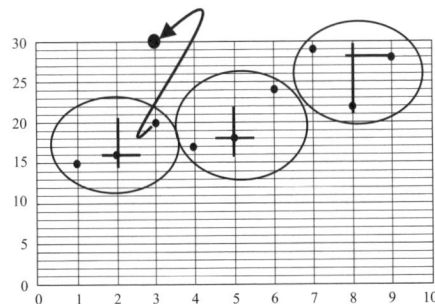

the position of the first summary point will not be altered. Thus, the slope will be the same as before.

Question 53 A

For that winter-time figure, 11 897.

$$\text{Seasonal index} = \frac{\text{observed value}}{\text{deseasonalised value}}$$

$$\therefore \quad 1.33 = \frac{11\,897}{\text{deseasonalised value}}$$

$$\therefore \quad \text{deseasonalised value} = \frac{11\,897}{1.33}$$

$$\approx \quad 8945$$

Question 54 C

For the two-term moving mean, do not write them all out. The full table as you go past May looks like this:

	April	May	June
	887	1139	1077
Mean		1013	1108

Which figure 'belongs' to May? Neither really. Find the mean – it comes to 1060.5.

Question 55 B

There is certainly an increasing trend which makes B a good answer so far. There is also seasonality. From the graph, it looks like the middle of each year is wine-drinking season...

Question 56 E

The y-intercept of this graph is 17. The gradient of this graph is negative, and by forming the largest triangle on it, the 'rise over run' for the line is:
$$\frac{(17-5)}{11} = \frac{12}{11} \approx 1.1$$

But the gradient is negative (look at it...) so we take -1.1. If we were using $y = ax + b$, then the equation would be $y = -1.1x + 17$. Using the correct words:
$$\text{rate} = -1.1 \times \text{year} + 17 \text{ or}$$
$$\text{rate} = 17 - 1.1 \times \text{year}$$

Solutions: A3

Question 57 D

Averaging four accidents at a time and writing the results in a third column gives the following – but note that you only really need to process up to 1997.

Year	Value	Smoothed
1993	13	
1994	7	
		8
1995	3	
		7.25
1996	9	
		7.5
1997	10	
		8.5
1998	8	
		7.75
1999	7	
		7.75
2000	6	
		8.5
2001	10	
2002	11	

Note that the answers are written 'between the years' (because 4 is an even number). Note also that the smoothed results start after the first two years and finish before the second last year.

For 1995, we average the results 8 and 7.25, giving 7.625. (Actually 8 and 7.25 are the only two numbers I should have actually calculated if I really wanted to conserve precious time).

Question 58 E

The sum of the quarterly indices must equal 4.
Thus, $1.05 + 0.84 + 0.92 +$ the missing number $= 4$.
$\therefore$ the missing number is 1.19.

Question 59 A

The word 'difference' here can be interpreted as meaning 'the vertical distance between the two graphs'. To answer this, I measured the distance between the graphs at each time period and plotted the result. Each 'star' represents one of the distance measurements.

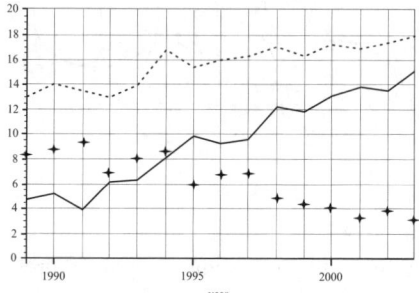

There is a definite decreasing trend in the stars. Actually, if I had wanted to save lots of time, I could have guessed the answer by looking at the 'closing up' between the two time series plots that we were given...

Question 60 D

The equation from the VCAA formula page which applies here is

$$\text{Seasonal index} = \frac{\text{actual figure}}{\text{deseasonalised figure}}$$

$\therefore \quad 1.28 = \dfrac{\text{actual figure}}{28\,098}$

$\therefore \quad 28\,098 = \dfrac{\text{actual figure}}{1.28}$

$\therefore \quad$ actual figure $= 28\,098 \times 1.28$
$ = 35\,965.44$
$ \approx 35\,965$ litres

Question 61 B

Draw a vertical line through the 'middle dot' as you move horizontally and draw a horizontal line through the 'middle dot' as you move vertically. The result looks like this:

Solutions: A3

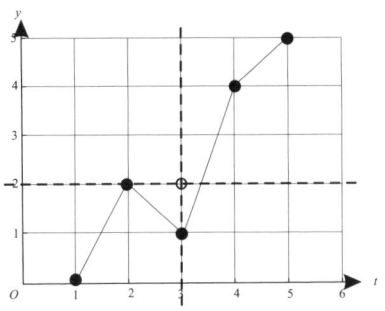

The coordinates of the median (it is circled) are (3, 2). Note: definitely *NOT* (3, 1).

Question 62 A

The fourth quarterly seasonal index:

Quarter	1	2	3	4
Seasonal index	1.3	0.9	0.7	1.1

A seasonal index of 1.1 means that an increase of $\frac{1.1}{1.0} \times \frac{100}{1} = 10\%$ will occur.

Question 63 B

$$\text{share price} = 1.24 + 0.06 \times \text{month}$$
$$\therefore \text{share price} = 1.24 + 0.06 \times 48$$
$$= \$4.12$$

Question 64 E

The vertical spread increases with time and there is an increasing line that fits well through the symmetrical centre of this graph. So, 'The share price shows an increasing linear trend with increasing variability' is perfect.

Question 65 C

The three percentage figures surrounding April are 35, 99 and 75. So the three mean smoothed average will be
$$\frac{35 + 99 + 75}{3}$$

$= 69.6666$
≈ 70 mm.

Question 66 D

The seasonal index values always add up to a number equal to the number of cells. Here, they must add up to 12. Because the numbers given add up to 11.02, the missing number must be 0.98.

Question 67 C

Use the equation from the formula sheet:
$$\text{seasonal index} = \frac{\text{actual figure}}{\text{deseasonalised figure}}$$
$$\therefore 0.94 = \frac{330}{\text{deseasonalised figure}}$$
$$\therefore \text{deseasonalised figure} = \frac{330}{0.94}$$
$$= 351.0638$$
$$\approx 351$$

Question 68 A

Given: deseasonalised number of unemployed $= 373.3 - 3.38 \times$ month number.
For June, the month number is 6. The equation predicts that the deseasonalised number of unemployed will be
$$373.3 - 3.38 \times \text{month number}$$
$$= 373.3 - 3.38 \times 6$$
$$= 353.02.$$
However, they are asking for the actual number of unemployed people. The seasonal index for June is 0.96:
$$\text{seasonal index} = \frac{\text{actual figure}}{\text{deseasonalised figure}}$$
$$\therefore 0.86 = \frac{\text{actual figure}}{353.02}$$
$$\therefore \text{actual figure} = 0.86 \times 353.02$$
$$= 303.5972$$
$$\approx 304$$

Solutions: A3

Question 69 E

There is definitely a straight line appearance about the pattern formed by most of the dots. Since the words 'skew' and 'correlation' make no sense here, seeing if seasonal variation exists is a good start. Yep, sure enough, those bumps do appear to be regularly spaced (12 months) apart don't they? Seasonal variation with an increasing trend.

Question 70 C

It would appear that there are 36 dots on the diagram. Group the first 12 and the last 12 points to find their summary points.

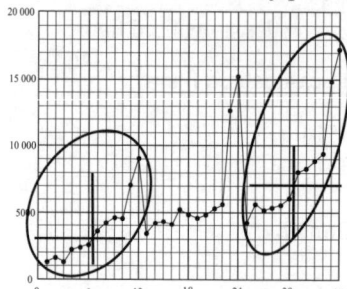

The summary points are about (6.5, 3000) and (30.5, 7000). Be sure there are 6 points either side of the median lines.
The gradient between these points is
$$\frac{rise}{run} = \frac{7000-3000}{30.5-6.5} = 166.66.$$
Since they do not ask for the equation of the actual line, just the gradient, there is no need to lift it $\frac{1}{3}$ of the way to a middle summary point. In fact there is no need for a middle summary point. The nearest alternative is **C**.

Question 71

First find the mean of the 12 months of data. The sum is 43 872 and so the mean is
$$\frac{43\,872}{12} \approx 3656.$$

Next, divide the value for September by this mean.
$$\frac{4597}{3656} = 1.2573 \approx 1.26.$$

166

Solutions: B1

Question 72 C

This is an AP with $a = 4$ and $d = 6$.
$$S_{10} = \frac{10}{2}(2(4) + 9(6))$$
$$= 5(8 + 54) = 310$$

Question 73 C

Given: $t_4 = a + 3d = 20$ and
$t_2 = a + d = 36$
Subtract these equations: $2d = -16$ and $d = -8$
Substitute this into the second equation:
$\therefore t_2 = a - 8 = 36$
$\therefore a = 44$

Question 74 C

The numbers are 47, 43, 39, ...
This is an AP with $a = 47$ and $d = -4$. The required term, t_{11}, is
$a + 10d = 47 + 10(-4) = 7$

Question 75 B

Try each in turn. Answer A is correct because you multiply by 0.1 each time. Answer B is terrible because if you choose a term and divide by the term before it, you get a different answer for each choice. C is tricky because $r = 1$. The other two, D and E, are definitely GPs.

Question 76 A

Three hours is 180 minutes. Therefore the amount of water that flows out is 180×120 litres.
The volume of the water remaining will be $18\,000 - 180 \times 120 = -3600$ litres. Ever tried drinking negative water? Dodgy stuff as it leaves you thirsty after you drink it. The answer is that the tank is empty.

Question 77 B

The increases are in a pattern:
8, 4, 2, ... This is a GP with $a = 8$ and $r = \frac{1}{2}$. The sum to infinity is
$$S_\infty = \frac{a}{1-r} = \frac{8}{1-\frac{1}{2}} = \frac{8}{\frac{1}{2}} = 16 \text{ cm.}$$
So it starts at 20 cm and will have a maximum height of $20 + 16 = 36$ cm.

Question 78 B

Terms in the sequence are 100, 40, 16, ...
This is a GP with $a = 100$ and $r = 0.4$. The terms will remain positive if you multiply by a positive number each time.

Question 79 E

Assuming that $t_1 = 20$, here are the terms that each answer generates:

A. gives 20, 0, 20, 0, ...
B. gives 20, 0, −20, −40, ...
C. gives 20, 10, 5, ...
D. gives 20, 10, 0, −10, ...
E. gives 20, 10, 20, 10, ... as required.

Question 80 A

The value of r is $\frac{7}{10}$ or $\frac{4.9}{7}$, whichever one you wish (they are the same). The fourth term is $4.9 \times \frac{7}{10} = \frac{34.3}{10} = 3.43$.

Question 81 B

In this geometric sequence −5.4, 1.8, −0.6, 0.2, the value of $a = -5.4$ and
$r = -\frac{1.8}{5.4} = -0.3333$. Thus $S_\infty = \frac{a}{1-r}$ is

Solutions: B1

$$= \frac{-5.4}{1-(-0.3333)}$$

$$= \frac{-5.4}{1.3333} \approx -4.05$$

Question 82   D

This is a geometric progression with $a = 600$. An increase of 8% p.a. means that $r = 1.08$. We want the third term, t_3.

$$t_n = ar^{n-1}$$
$$\therefore \quad t_3 = ar^2$$
$$= 600 \times 1.08^2$$
$$= 699.84$$

The number of goats on the farm will be closest to 700. (Don't be worried about the missing 0.16 of a goat. Maybe they become woollier as winter approaches.)

Question 83   D

The progression is 5, 7, 9, ... It is an A.P. with $a = 5$, $d = 2$. We want

$$t_{12} = a + (n-1)d$$
$$= 5 + 11(2)$$
$$= 27 \text{ laps}$$

Question 84   E

To count the number of laps, add the terms 5, 7, 9, ... up to the 20th term and multiply by 7, since each day is repeated 7 times (you need to read this question carefully...).

$$\therefore \quad S_n = \frac{n}{2}(2a + (n-1)d) \times 7$$
$$\therefore \quad S_{20} = \frac{20}{2}(2(5) + (20-1)2) \times 7$$
$$= 10(10 + 38) \times 7$$
$$= 3360 \text{ laps}$$

Question 85   D

Given: $w_{n+1} = 4w_n + 2$
$\therefore \quad w_3 = 4w_2 + 2$

$$= 4 \times 10 + 2$$
$$= 42$$

Thus the fourth term is
$\therefore \quad w_4 = 4w_3 + 2$
$$= 4 \times 42 + 2$$
$$= 170$$

Note that we do not need the first term.

Question 86   E

There are several approaches to this question. Here is one of them:

$$t_1 = 15$$
$\therefore \quad t_2 = t_1 - 5 = 15 - 5 = 10$
$\therefore \quad t_3 = t_2 - 5 = 10 - 5 = 5$

This would appear to be an AP with $a = 15$ and $d = -5$.

$\therefore \quad t_n = a + (n-1)d$
$$= 15 + (n-1)(-5)$$
$$= 15 - 5n + 5$$
$$= 20 - 5n$$

(Try it: put $n = 1$, $n = 2$ and so on into this answer.)

Question 87   C

This is an arithmetic progression since 6 is being added each time. It is probably quickest to simply list more terms until 23 is exceeded (it is the biggest). Thus,
-5, 1, 7, 13, 19, 25, ... Note that 19 appears.

Question 88   D

A comment on each one.
1.1, 1.11, 1.111, 1.1111 ... Not geometric.
3, 6, 12, 24... Geometric with $r = 2$.
$\frac{1}{2}, \frac{1}{4}, \frac{1}{8}, \frac{1}{16}$ ... Geometric with $r = \frac{1}{2}$.
27, 9, 3, 1 ... Geometric with $r = \frac{1}{3}$.

So, there are three.

Solutions: B1

Question 89 D

Let x be the amount to be paid this year. Then it is true that
$$\frac{1200}{x} = \frac{3}{5}$$
$\therefore\ 3x = 5 \times 1200$
$\therefore\ x = \$2000$.

Question 90 C

The number of blocks: 3, 5, 7, ...This is an arithmetic progression with $a = 3$ and $d = 2$. We wish to have a sum of 50 at most.

$S_n = \dfrac{n}{2}(2a + (n-1)d)$

$= \dfrac{n}{2}(2(3) + (n-1)(2))$

$= \dfrac{n}{2}(2n + 6 - 2)$

$= \dfrac{n}{2}(2n + 4)$

Try each number from 1 onwards for n until you exceed 50. You could use a table or simply substitute n values. I found that $n = 6$ gives a sum of 48 but $n = 7$ gives a sum of 63.

Question 91 E

A geometric progression with $S_\infty = 80$ is presented to us. We are told that $a = 80$ and $r = 0.75$.

$S_\infty = \dfrac{a}{(1-r)}$

$= \dfrac{80}{(1-0.75)}$

$= \dfrac{80}{0.25}$

$= 320$

Question 92 B

Let t_n be the term we are currently looking at. To get the next term, multiply this by 5, giving $5t_n$, and then subtract 2. This gives $t_{n+1} = 5t_n - 2$. Note that the next term is called t_{n+1} and that the first term is 3.

Question 93 A

The graph shows that the terms are getting smaller. That is, they are becoming closer to zero. But it is really important to also see that the terms jump from $+$ to $-$ values making r a negative number. Yes, r is small but it is negative. Thus, $-1 < r < 0$ is perfect.

Question 94 C

Interpret the equation $t_{n+1} = at_n + b$. It means that you take the previous term, multiply by a number and then add another number to give the next term. It is a combination of an A.P. and a G.P. The sequence is purely arithmetic only if $a = 1$ because to get the next term, only b will be added each time. Note that we don't care what b is.

Question 95 B

It pays to rearrange the equation so that 'the previous term' is the subject:

$P_n = 2P_{n-1} - 200$

$\therefore\ 2P_{n-1} = P_n + 200$

$\therefore\ P_{n-1} = \dfrac{1}{2}P_n + 100$

To find P_5,

$P_5 = \dfrac{1}{2}P_6 + 100$

$= \dfrac{1}{2} \times 1000 + 100$

$= 600$.

You can't stop there! We must find P_4...

$P_4 = \dfrac{1}{2}P_5 + 100$

$= \dfrac{1}{2} \times 600 + 100$

$= 400$.

Solutions: B1

Question 96 D

Take any term and *subtract* the term before it. The result should be the same each time. This is only true in D where this subtraction gives 3 every time. The others give a different answer each time.

Question 97 C

Dividing x by 6 must give the same as dividing 54 by x:
$$\therefore \quad \frac{x}{6} = \frac{54}{x}$$
$$\therefore \quad x^2 = 54 \times 6$$
Note that some students will go straight to this idea using the concept of 'geometric mean'.
$$\therefore \quad x = \sqrt{324}$$
$$\therefore \quad x = 18$$

Question 98 C

To increase a number by 84%, multiply it by 1.84. Now combine this with the sale of 40. The number of sheep will follow this pattern: $50, 50 \times 1.84 - 40, \ldots$
That is, $50, 92 - 40, \ldots$
At the start of the second year, she has $92 - 40 = 52$ sheep.

Question 99 A

The required difference equation will take last year's count (S_n), multiply it by 1.84 and then take away the 40 that are sold.
$$\therefore \quad S_{n+1} = 1.84 S_n - 40.$$ The initial number, $S_1 = 50$ should be stated.

Question 100 D

$$f_{n+1} - f_n = 5$$
$$f_1 = -1 \text{ (given)}$$
$$\therefore \quad f_2 - f_1 = 5$$
$$\therefore \quad f_2 - (-1) = 5$$
$$\therefore \quad f_2 = 5 - 1 = 4$$
$$\therefore \quad f_3 - f_2 = 5$$
$$\therefore \quad f_2 - (4) = 5$$
$$\therefore \quad f_3 = 5 + 4 = 9$$
That is, $-1, 4, 9, \ldots$

Question 101 E

The lengths form a geometric progression with $a = 12.0$ cm and $r = \left(1 + \dfrac{3}{100}\right) = 1.03$.

Now, the expression 'after 14 days' means that we want the 15th term. Why? If, for example, it said 'after 1 day', you would be looking at the second term.
That is, $t_{15} = ar^{15-1}$
$\phantom{That is, t_{15}} = 12.0 \times 1.03^{14}$
$\phantom{That is, t_{15}} = 18.1507$
$\phantom{That is, t_{15}} \approx 18.2$ cm

Question 102 B

In this question, you need to look closely at each alternative answer. The dots on the graph form a curve. This allows us to reject A, C and D since they add a constant to the previous term, resulting in a straight line. Now, term 1 is about 4. From alternative E, term 2 would be $4 \times 3 = 12$ which is not what is shown on the graph. However, using alternative B, term 2 would be $2 \times 4 + 1 = 9$ which looks just right…

Question 103 D

The number of *new* stamps in Paula's stamp collection form this progression:
$1, 2, (1 + 2 = 3), (3 + 2 = 5), \ldots$
$= 1, 2, 3, 5, 8, 13, \ldots$
Add the first 5 terms:
$1 + 2 + 3 + 5 + 8 = 19$ stamps.

Question 104 E

Start by putting the increases in a line, using kilograms as the unit:
$0.4, 0.38, 0.361, \ldots$ Now look for a pattern. They are not decreasing by a constant, so they are not in an arithmetic sequence. If,

however, you divide a term by the previous term, you get 0.95 each time. This is a Geometric Progression. His 'eventual body weight' is found by finding the sum to infinity of these terms. $a = 0.400$ and $r = 0.950$. $S_\infty = \dfrac{a}{1-r} = \dfrac{0.4}{1-0.95} = 8$ kg.
Now add his initial weight of 73.4 giving 81.4 kg. He won't worry about his weight by then. He will be too old…

$= 680 \times 0.96^5$
$= 554$ waterfowl.

Question 109 E

We know that $W_1 = 680$.
The second term is $W_2 = 0.96 \times W_1$.
The third term is $W_3 = 0.96 \times W_2$ and so on.
The required difference equation is
$W_{n+1} = 0.96 W_n$

Question 105 B

To find the common ratio of a geometric sequence, choose any term (not the first one) and divide by the term before it. In this case, $\dfrac{6}{24} = 0.25$.

Question 106 C

Consider the fact that data forms the first three terms of an arithmetic sequence. The terms are 15, X, 29, …
$t_3 = a + 2d$
∴ $29 = 15 + 2d$
∴ $d = 7$
∴ $t_4 = a + 3d$
$= 15 + 3(7)$
$= 36$

Question 107 B

$t_{n+1} = at_n + 6$
∴ $t_2 = a(t_1) + 6$
∴ $21 = a(5) + 6$
∴ $15 = a(5)$
∴ $a = 3$

Question 108 C

The numbers of waterfowl form a geometric progression.
$680, 680 \times 0.96, 680 \times 0.96^2, …$
Five years later, the required number will be the sixth term of this progression.
$t_5 = ar^{6-1}$

Question 110 B

The terms in the progression involved here are
$20, 19, 18, …$
This is an arithmetic progression with $a = 20, d = -1$
Logically, the stack is complete when the number of cans in a row is 1.
$t_n = a + (n-1)d$
∴ $1 = 20 + (n-1)(-1)$
∴ $(n-1)(-1) = -19$
∴ $n - 1 = 19$
∴ $n = 20$
The total number of cans will be
$S_n = \dfrac{n}{2}(2a + (n-1)d)$
$= \dfrac{20}{2}(2(20) + 19(-1))$
$= 10(40 - 19)$
$= 210$ cans

Question 111 E

If the common ratio of this sequence is negative, then the dots would alternate between being above the x-axis and being below. Only **D** and **E** support this. If the first term is positive, then the first dot will be above the horizontal axis. Hence alternative **E**.

Solutions: B1

Question 112 D

The first four terms of a sequence are 12, 18. 30 and 54. Start by computing the value of the second term of each sequence by using the difference equations. Darn, they all give $t_2 = 18$. Ok, so try calculating t_3 from $t_2 = 18$. **A** fails (24), **B** fails (27), **C** fails (21), **D** and **E** both correctly give 30. Don't give up, try computing t_4. **D** works (54) but **E** fails (40).

Question 113 A

The total amount of forest destroyed by the eruption will be the sum of the first 9 terms of the progression,
15, 13.5, 12.15, ...
If you take one of those terms and divide by the term before it, you get 0.9, making it a geometric progression.
The sum of the first 9 terms is

$$S_n = \frac{a(1-r^n)}{1-r}$$

$$= \frac{15(1-0.9^{14})}{1-0.9}$$

$$= 115.6848$$

$$\approx 116 \text{ square kilometers.}$$

Solutions: B2

Question 114 E

$$\sin(\angle BAC) = \frac{\text{opposite}}{\text{hypotenuse}}$$
$$= \frac{16}{25}$$

On my calculator, I press 2nd sin (16/25) and the result is 39.79°, closest to 40°.

Question 115 B

Since X is due west of Z, we can add some lines to our diagram:

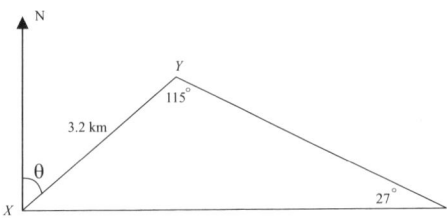

Angle YXZ must be $180 - 115 - 27 = 38°$. Thus, the angle marked θ must be $90 - 38 = 52°$. The bearing is 052°.

Question 116 C

Using the sine rule:
$$\frac{XZ}{\sin 115°} = \frac{3.2}{\sin 27°}$$
$$\therefore \quad XZ = \frac{3.2 \times \sin 115°}{\sin 27°}$$
$$= 6.388$$
$$\approx 6.4 \text{ km.}$$

Question 117 E

The two triangles may look like this:

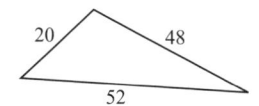

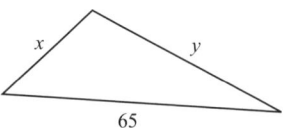

We will use the algebra associated with similar triangles. First, to find x:
$$\frac{65}{52} = \frac{x}{20}$$
$$\therefore \quad x = \frac{65 \times 20}{52}$$
$$= 25 \text{ cm}$$

to find y:
$$\frac{65}{52} = \frac{y}{48}$$
$$\therefore \quad y = \frac{65 \times 48}{52}$$
$$= 60 \text{ cm}$$

Thus, the perimeter will be $65 + 25 + 60$ which is 150 cm

Question 118 A

V is at altitude 450 m. U is at altitude 300 m. The vertical distance between them is 150 m. We are told in the question that the horizontal distance is 200 m.
Thus the $\frac{\text{rise}}{\text{run}}$ is $\frac{150}{200} = 0.75$.

Question 119 D

Using the cosine rule with ΔFGH is **not** a good idea. When the unknown length is not opposite the known angle, things become tricky.
Here is ΔFHI redrawn with a little more information:

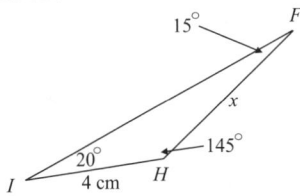

173

Solutions: B2

$$\frac{4}{\sin 15°} = \frac{x}{\sin 20°}$$
∴ $x = \frac{4\sin 20°}{\sin 15°}$

Question 120 A

First, the length of *AD* is needed.
Pythagoras' theorem on the base:
$AD = \sqrt{100^2 + 50^2} \approx 118.03$
Next, draw $\triangle ABD$:

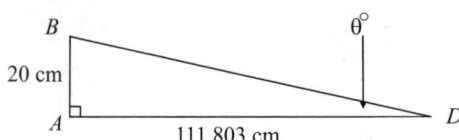

$\tan \theta = \dfrac{20}{111.803}$
∴ $\theta = 10.142° \approx 10.1°$

Question 121 C

The area of such a triangle is given by
$A = \dfrac{1}{2}ab\sin(c)$
$= \dfrac{1}{2}(5)(10)\sin(25°)$
$= 10.5654$
≈ 10.6 cm.

Question 122 C

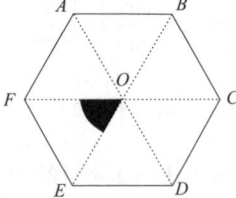

The centre of a regular hexagon is divided into 6 parts.
The size of obtuse angle *FOE* therefore is $\dfrac{360}{6} = 60°$.
Double this to give angle *FOD* (120°).

Question 123 E

Mark the point under that right-angle as point *M*.

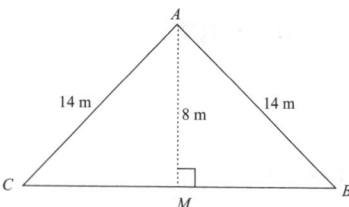

Use Pythagoras on triangle *CMA*.
$CM^2 + 8^2 = 14^2$
∴ $CM^2 = 14^2 - 8^2$
∴ $CM = 11.489$
≈ 11.5 m.
Oops - I forgot to double it! The length all the way from *C* to *B* is $11.5 \times 2 = 23$ m.

Question 124 B

Look at triangle *OAB*. It has two measurements plus one angle. It's time for the cosine rule...
$x^2 = 6^2 + 4^2 - 2(6)(4)\cos(30°)$
$ = 52 - 48\cos(30°)$
(Don't subtract 48 from 52 at this stage.)
$ = 10.4307$
∴ $x = \sqrt{11.489} = 3.2296$
≈ 3.2 km.

Question 125 E

Definitely, a diagram with all the information is called for at this stage:

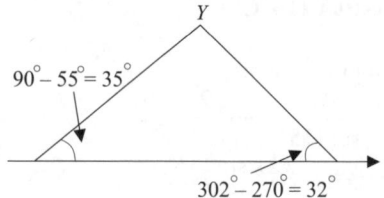

The angles inside the triangle are needed:

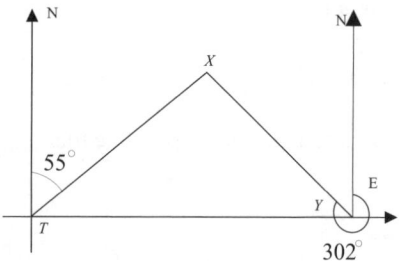

Solutions: B2

The angle at *Y* is found by subtracting the interior angles from 180°.
$$180° - 32° - 35° = 113°.$$

Question 126 B

A direct application of the sine rule is called for:
$$\frac{20}{\sin \alpha°} = \frac{15}{\sin \beta°}$$
But we are told that $\sin \alpha° = 0.8$.
$$\therefore \quad \frac{20}{0.8} = \frac{15}{\sin \beta°}$$
$$\therefore \quad 20\sin \beta° = 0.8 \times 15$$
$$\therefore \quad \sin \beta° = \frac{0.8 \times 15}{20} = 0.6$$

Question 127 A

Here is a drawing of the 'floor':

We want to find the length from *H* to *M*:
$$HM^2 = 4^2 + 2^2$$
$$\therefore \quad HM = \sqrt{20}$$
Now for the interior triangle ΔEHM:
$$\tan(EMH) = \frac{EH}{HM}$$
$$= \frac{4}{\sqrt{20}}$$
$$= 0.8944$$
Press 2nd tan(0.8944). The answer is 41.8°.

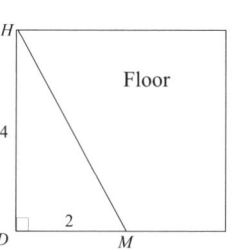

Question 128 B

From the triangle, $\sin \theta = \frac{142}{215}$
Pressing 2nd sin then that fraction (making sure you are in degrees mode) gives $41.335° \approx 41°$.

Question 129 D

There are no right angles here, but there are two given angles. This points to using the sine rule.
$$\frac{45}{\sin 54°} = \frac{RT}{\sin 108°}$$
$$\therefore \quad RT = \frac{45 \times \sin 108°}{\sin 54°}$$
$$= 52.9007$$
$$\approx 53 \text{ cm}$$

Question 130 A

Point *A* is on the 300 metre contour. Point *B* is on the 200 metre contour. Therefore the difference in height is 100 m.

Question 131 C

There are a number of ways of approaching this.

Here, I will find the area of the front and multiply by 24 (the depth).

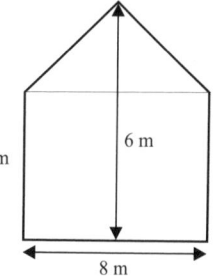

The square (underneath) has area $8 \times 4 = 32 \text{ m}^2$ and the triangle (at the top) has area $= \frac{1}{2}$ base × height
$$= \frac{1}{2} \times 8 \times (6-4)$$
$$= 8 \text{ m}^2$$
Thus the whole front has area 40 m².
$$\therefore \quad \text{the volume is } 40 \times 24 = 960 \text{ m}^2.$$

Question 132 C

To use Heron's formula, start by adding up the side lengths: $36 + 58 + 42 = 136$. Now halve it: 68. This is '*s*'.
Heron:
$$A = \sqrt{s(s-a)(s-b)(s-c)}$$
$$= \sqrt{68(68-36)(68-58)(68-42)}$$

Solutions: B2

Wooah! Don't go any further. Check the form of the answers... it is alternative **C** already.

Question 133 D

We will need the dimensions of the box. The measurements are 1.5, 2 and h cm. Multiplying them together gives the volume, which is known to be 6 m³.
$\therefore \quad 1.5 \times 2 \times h = 6$
$\therefore \quad h = \dfrac{6}{1.5 \times 2}$
$= \quad 2$ cm.

The total surface area will be
$2(1.5 \times 2 + 1.5 \times 2 + 2 \times 2)$
$= \quad 2(3 + 3 + 4)$
$= \quad 20$ m².

Question 134 E

This is a similar triangle problem. Here it is a good idea to separate our two similar triangles as shown in this diagram:

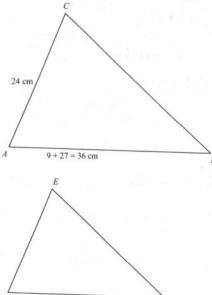

From these,
$\dfrac{CA}{ED} = \dfrac{9 + 27}{27}$
$\therefore \quad \dfrac{24}{ED} = \dfrac{36}{27}$
$\therefore \quad 36 \times ED = 24 \times 27$
$\therefore \quad ED = \dfrac{24 \times 27}{36}$
$= \quad 18$ cm

Question 135 A

The diagram needs some work. First, mark the radius in a second place, as shown.

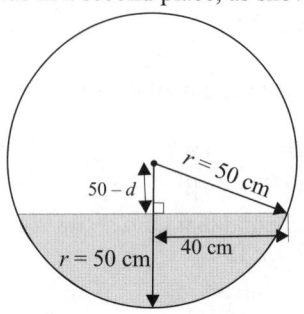

Next, create a right-angled triangle in the white space, with the height being the difference between 50 and the depth.
From the right-angled triangle,
$r^2 = 40^2 + (r - d)^2$
$\therefore \quad 50^2 = 40^2 + (50 - d)^2$

Now, you have two choices. Either try quadratic algebra or try using their alternatives in this equation until you find the one that works.
$\therefore \quad (50 - d)^2 = 2500 - 1600$
$\therefore \quad (50 - d)^2 = 900$
$\therefore \quad 50 - d = 30$ (a positive number)
$\therefore \quad d = 20$

Question 136 D

For this question, it is important to construct a diagram with the bearing information carefully constructed on it. Note that the triangle is isosceles. (That is why the letter θ is written twice.)

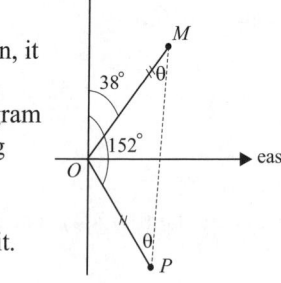

From the diagram, the triangle has angles θ°, θ° and (152 − 38)°. But they add up to 180°:
$\theta + \theta + 114 = 180$
$\therefore \quad 2\theta = 66$
$\therefore \quad \theta = 33°$

176

Solutions: B2

Now draw the diagram again to show this new information.

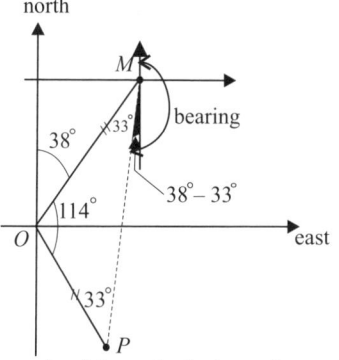

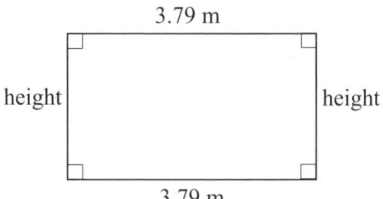

Check each of the calculations shown here and locate the unknown:

The bearing is $180 + (38 - 33)$
$= 185°$
This is 'between $180°$ and $270°$'.

Question 137 A

The triangle is a right-angled triangle, so conventional trigonometry is used.

$$\cos \theta° = \frac{\text{adjacent}}{\text{hypotenuse}}$$

∴ $\cos(\theta°) = \frac{6}{10}$.

Question 138 B

Once again, this problem uses conventional trigonometry.

$$\tan \theta° = \frac{\text{opposite}}{\text{adjacent}}$$

∴ $\tan(37°) = \frac{BC}{2200}$.

∴ $BC = \tan(37°) \times 2200$
$= 1657.8189$
≈ 1658 m.

Question 139 B

Here is a drawing of the rectangle:

Since the perimeter is 24.50 m, then
$2 \times 3.79 + 2 \times (\text{height}) = 24.5$
∴ $3.79 + (\text{height}) = 12.25$
∴ height = 8.46 m
The length of the diagonal can now be found by using Pythagoras' theorem.
$D = \sqrt{8.46^2 + 3.79^2}$
$= 9.27015 \approx 9.3$ m.

Question 140 A

This beam is in the shape of a prism. The volume of a prism is found by multiplying the area of the cross-section by the length. In this case, the volume is
$0.048 \times 12 = 0.576$ m^2.

Question 141 B

Comparing a map to the real world, the ratio of the areas is equal to the square of the ratio of lengths. Thus,

$$\frac{A_2}{A_1} = \left(\frac{L_2}{L_1}\right)^2$$

∴ $\frac{4000}{10} = \left(\frac{x}{1}\right)^2$

∴ $x^2 = 400$
∴ $x = 20$ m.

Question 142 D

The total surface area of the remaining shape consists of three parts.
Underneath there is a circle with area $\pi r^2 = \pi \times 20^2$ m^2.
The curved surface around outside has area $2\pi rh = 2 \times \pi \times 20 \times 30 = 1200\pi$ m^2.
At the top, inside the shape, is half of a sphere. Its surface area is

$$\frac{1}{2} \times 4\pi \times r^2 = 2\pi \times 20^2 \text{ m}^2.$$

Therefore, the total surface area is equal to
$400\pi + 1200\pi + 800\pi$
$= 2400\pi$
$= 7539.822$
$\approx 7540 \text{ m}^2$

Question 143 C

Here is a diagram of the shapes;

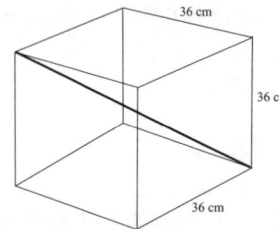

The length of the metal rod can be found by using Pythagoras' theorem is three dimensions.

$L = \sqrt{36^2 + 36^2 + 36^2}$
$= 62.3538$
$\approx 62 \text{ cm}$

Question 144 D

This is a clever question. For a start, only half of the contour map appears in the alternatives. Next, X (250 m) is higher than Y (50 m). That eliminates **A**, **B** and **D**. Now look at the two contour lines nearest X. They are close together which means it is steep there. Only **D** shows that feature.

Question 145 E

Definitely, a diagram is called for here.

It would appear that Graeme used the cosine rule.

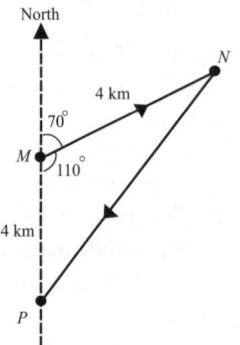

$a^2 = b^2 + c^2 - 2bc\cos(A)$
$\therefore \quad NP^2 = 4^2 + 4^2 - 2 \times 4 \times 4 \times \cos 110°$
$= \sqrt{16 + 16 - 2 \times 4 \times 4 \times \cos 110°}$

Shelley has recognized that the triangle is isosceles. dividing it down the centre provides right-angled triangles, with x being half the answer:

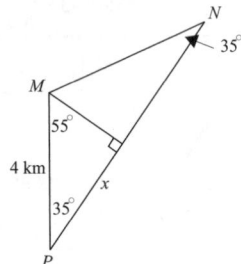

From this triangle, $\cos 35° = \dfrac{x}{4}$

$\therefore \quad x = 4\cos(35°)$
$\therefore \quad NP = 2 \times 4\cos(35°)$

Tran has seen that the angles in the triangle are 35°, 35° and 110°. Using the sine rule,

$$\frac{4}{\sin 35°} = \frac{NP}{\sin 110°}$$

$\therefore \quad NP = \dfrac{4 \times \sin 110°}{\sin 35°}$

All three people are correct.

Solutions: B3

Question 146 D

There are two telephone calls. The 90-second call costs $1 and the 30-second call costs $0.50. The total is $1.50. Note that the open circle above the 30-second line is not used. The coloured-in circle ($0.50) is used.

Question 147 B

If you substitute the point (2, 1) into the equation $y = 3x + c$, you obtain a statement of truth (because the point lies on the line).
$$y = 3x + c$$
$$\therefore \quad 1 = 3(2) + c$$
$$\therefore \quad 1 = 6 + c$$
$$\therefore \quad c = -5$$

Question 148 E

$y + 8 = 0$ is the same as $y = -8$ and $x - 12 = 0$ is the same as $x = 12$.
Thus they intersect at $x = 12$ and $y = -8$, or better, (12, –8). Now *there* is an unusual question…

Question 149 C

The speed of the car is measured by the gradient of the distance-time graph. In this version of the diagram, three triangles are shown. From left to right the gradients are calculated:

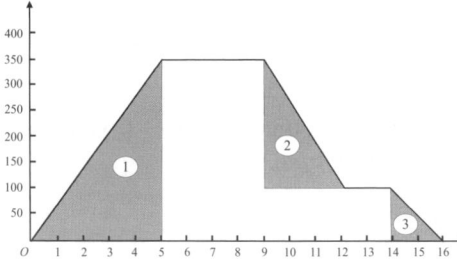

The first: $\dfrac{350}{5} = 70\,\text{km/h}$

The second: $\dfrac{100 - 350}{12 - 9} = \dfrac{-250}{3}$
$= -83.3\,\text{km/h}$

The third: $\dfrac{-100}{16 - 14} = \dfrac{-100}{2} = -50\,\text{km/h}$

We ignore the negative signs since the question talks about *speeds*. The negatives are irrelevant.
So the speed is greatest in 83.3 km/h in the time between 9 and 12 hours.

Question 150 D

From time = 0 to time = 5, the car travelled 350 km. From time = 5 until time = 9, the car was not moving. From time = 9 until time = 12, the car travelled 250 km (backwards). The total is 600 km.

Question 151 E

We can form two simultaneous equations from the data:
When $x = 4$, $C = 320$
When $x = 6$, $C = 450$
Substituting these into the equation $C = ax + b$ (and using the one with larger numbers first) we obtain:
$$450 = a(6) + b \quad \text{and}$$
$$320 = a(4) + b$$
or
$$450 = 6a + b \quad \text{and}$$
$$320 = 4a + b \,...1.$$
subtract:
$$130 = 2a$$
$$\therefore \quad a = 65$$
Substitute this back into equation 1:
$$320 = 4(65) + b$$
$$\therefore \quad b = 320 - 260$$
$$= 60$$
This means that $C = 65x + 60$
So when $x = 1$, $C = 65(1) + 60$
$$= \$125$$

Solutions: B3

Question 152 C

We will list the corner ordered pairs and then find the value of $y - 2x + 20$ at each point.

Point	$y - 2x + 20$
1 (2, 9)	25
2 (4, 11)	23
3 (6, 10)	18
4 (6, 1)	9

That is, the maximum value of $y - 2x + 20$ occurs at (2, 9) with a value of 25.

Question 153 B

Start by looking at $3x + 4y = 12$. Its x-intercept will be 4 and its y-intercept will be 3. This matches alternatives A, B and C but rejects D and E.
If you substitute (0, 0) into $3x + 4y \leq 12$, you get $0 \leq 12$ which is true. Therefore the shaded feasible region is under the line $3x + 4y = 12$, making A and B candidates.
Now try $x - y \leq 1$. Substituting (0, 0) gives $0 \leq 1$ which is true. However, the shaded (feasible region) is *under* the line $y = x$ and (0, 0) is above it so it should have given 'false'. That makes B the only remaining alternative.
There are other ways to solve this question.

Question 154 A

Let's find k in the expression $y = kx^3$.
From the original graph, when $x^3 = 8$, we know that $y = 1$. That is why (8, 1) appears on the graph whose horizontal axis is x^3.
$\therefore \quad y = kx^3$
$\therefore \quad 1 = k(8)$ so that $k = \dfrac{1}{8}$
$\therefore \quad y = \dfrac{1}{8}x^3$. On the lower graph set, we want the ordered pair shown to 'find' that equation. Try $x = 2, y = 1$.

Question 155 A

The question is asking 'where is the graph steepest and sloping to the right?'. The answer is circled here:

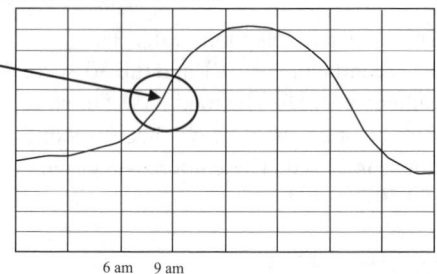

It is in the 6 am – 9 am section.

Question 156 E

We wish to solve $y = -5$ and $y = -x + 5$ simultaneously. Using algebra is best here:
$\qquad -x + 5 = -5$
$\therefore \qquad -x = -10$
$\therefore \qquad x = 10$
Substitute this into either equation. Probably $y = -5$ is the easiest…
Answer: (10, −5).

Question 157 C

Watch out for the word 'not'!

A. Substitute $x = 4, y = 7$ into $7x - 4y = 0$. Yes, it is true that $28 - 28 = 0$.
B. Substitute $x = 3, y = 5$ into $7x - 4y \geq 0$. Is $21 - 15 \geq 0$? Yes.
C. The line with equation $3x + 5y = 0$ can be rearranged into $y = -\dfrac{3}{5}x$. Ha! This one has a negative gradient. 'Not true!'. Why go any further?
D. If you substitute $x = 0, y = 0$ into both $7x - 4y = 0$ and $3x + 5y = 0$, you get statements of truth.

180

Solutions: B3

E. $7x - 4y = 0$ can be written as $y = \dfrac{7}{4}x$.

It has a positive gradient so yes it is true, y increases as x increases.

Question 158 C

Rearrange the equation $3x + 2y = 4k$

$3x + 2y = 4k$

$\therefore \quad 2y = -3x + 4k$

$\therefore \quad y = -\dfrac{3x}{2} + 2k$

But we can see from the graph that the y-intercept is 12. Therefore,

$2k = 12$

$\therefore \quad k = 6$

Question 159 D

The answer will be a linear equation with 50 as the fixed part and $65n$ as the variable part. Thus, $65n + 50$.

Question 160 E

Simultaneous equations are needed here. Let I be the number of ice creams and D be the number of drinks:

$4I + 3D = 21.40$
$5I + 2D = 20.80$

Multiply the top line by 2 and the bottom line by -3:

$8I + 6D = 42.80$
$-15I - 6D = -62.40$

Add the equations:

$-7I = -19.60$

$\therefore \quad I = 2.80$

Substitute this back into the first equation:

$4(2.80) + 3D = 21.40$

$\therefore \quad 3D = 21.40 - 11.20$

$\therefore \quad D = \dfrac{10.20}{3} = 3.40$

The price of one drink is $3.40.

Question 161 B

If the axes are renamed x and y, then the graph will have equation $y = ax + 0$ (because the y-intercept is 0). The gradient of the line is found by considering $(0, 0)$ and $(2, 4)$. $\dfrac{4-0}{2-0} = 2$. Hence the equation $y = 2x$. Now replace x with a^2 and y with b. Answer: $b = 2a^2$.

Question 162 E

We will list the corner ordered pairs and then find the value of $P = 4x - 3y$ at each point.

Point	$P = 4x - 3y$
1 (0, 0)	0
2 (0, 100)	–300
3 (50, 100)	–100
4 (90, 60)	180
5 (120, 0)	480

That is, the maximum value of $P = 4x - 3y$ occurs at $(120, 0)$ with a value of 480.

Question 163 A

The graph of $y = \dfrac{1}{x^2}$ provides us with the shape seen in alternatives A, B and C only. We are told in the text of the question that when $d = 50$, $I = 20$. Unfortunately that does not mean that $(20, 50)$ is on the graph. Part B looks good at first, but unfortunately, the 50 and the 20 are the wrong way around.

First find k:

$I = \dfrac{k}{d^2}$

Use $(50, 20)$:

$\therefore \quad 20 = \dfrac{k}{50^2}$

$\therefore \quad k = 20 \times 2500$

$\quad = 50\,000$

$\quad I = \dfrac{50\,000}{d^2}$

Solutions: B3

When $d = 50$,
$$I = \frac{50000}{50^2}$$
$$= 2000$$
Thus, look for (5, 2000) on the graph.
Strange how often the last multiple choice question of an exam is tricky, isn't it?

Question 164 D

A horizontal line will have the equation '$y = $ a number'.
At the point (1, 2), the value of y is 2.
Thus, the line is $y = 2$.

Question 165 E

It is a good idea to construct a horizontal line through the depth value of 10. Note there is a 'split scale' used here. Some other depth values have been put on the vertical axis for clarity:

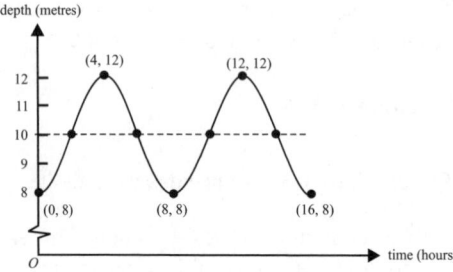

There are 4 intersections with the graph, hence 4 times when the height is 10 metres.

Question 166 C

At the bend in the middle of the graph, the time is 4.5 hours and the cylinder weighs 220 grams. That is when the gas runs out as the mass stays the same after that.

Question 167 A

After time $t = 4.5$ hours, the line is horizontal. Such a line will have equation $M = 220$. So far, that means alternatives A and E. So what is the difference between these possibilities? It is the gradient of the straight line. Try substituting $t = 4.5$ into each equation. For alternative A:
$$M = 332.5 - 25t$$
$$= 332.5 - 25 \times 4.5$$
$$= 220$$
For alternative E:
$$M = 332.5 - 12.5t$$
$$= 332.5 - 12.5 \times 4.5$$
$$= 230$$
Clearly, alternative A gives the correct answer.

Question 168 B

Watch out for the expression '**not** true'.
$$12x - 4y = 0$$
$$\therefore \quad 4y = 12x$$
$$\therefore \quad y = 3x$$
This is a straight line of gradient 3 passing through the origin. Alternative B is wrong. Note that C and D are both correct. It is worth checking that now – but not during the exam because it will use up valuable time.

Question 169 C

The question is asking this: 'in which of these alternatives does substituting $x = 2$ and $y = -2$ give a true statement? Try each in turn.
A. $x - y = 0$ $2 - (-2) = 2 + 4 = 4$ No.
B. $2x + 2y = 8$ $4 + (-2) = 4 - 2 = 2$ No.
C. $2x + 2y = 0$ $4 + (-4) = 4 - 4 = 0$ Yes!
D. $2x - 2y = 4$ $4 - (-4) = 4 + 4 = 8$ No.
E. $2x - 2y = 0$ $4 - (-4) = 4 + 4 = 8$ No

Question 170 E

The constraints here are:
$y \le 60$
$x \le 40$
$x + y \le 80$
and the usual $x \ge 0$, $y \ge 0$.

Solutions: B3

Given that x is for cows and y is for sheep, the first two alternatives are nonsense. The next two are reversed and the last one is exactly the same as $x + y \leq 80$.

Question 171 A

Use some algebra. Let x be the number of frying pans and let C be the cost
$\therefore \quad C = 400 + 50x$.

x	$\$C$	Answer
10	900	A, B
15	1150	C, D
20	1400	E

To 'break even', the money generated in the answers must match that. They are $900, $450, $900, $45 and $100 respectively. Answer A matches the cost.

Question 172 C

All of the alternatives obey the constraints $x \geq 0$ and $y \geq 0$. Test the others for being true (T) or false (F). Two Ts are needed.

Alternative	$x + y \leq 9$	$y \leq \dfrac{1}{2}x$	T/F
A	$4 + 4 \leq 9$	$4 \leq \dfrac{1}{2} \cdot 4$	TF
B	$5 + 3 \leq 9$	$3 \leq \dfrac{1}{2} \times 5$	TF
C	$6 + 2 \leq 9$	$2 \leq \dfrac{1}{2} \times 6$	TT
D	$6 + 4 \leq 9$	$4 \leq \dfrac{1}{2} \times 4$	FF
E	$7 + 3 \leq 9$	$3 \leq \dfrac{1}{2} \times 7$	FT

Alternative C produces a true result for all *four* constraints.

Question 173 D

Take two points on the graph. (0, 0) and (2, 1). The slope is
$$\frac{y_2 - y_1}{x_2 - x_1}$$
$$= \frac{1 - 0}{2 - 0}$$
$$= \frac{1}{2}.$$

Question 174 C

Here is a table with hours of work in one column, and the result of substituting those numbers into the equation $C = 60 + 55n$ in the other column.

Hours	Fee
1	$115
2	$170
8	$500
10	$610

Now compare each line in the table with the alternatives. $500 for 8 hours work appears there.

Question 175 E

The minimum temperature refers to the lowest point on the graph. A reminder – maximum and minimum points occur at end-points of a graph as well as on turning points. Here, the lowest point is on the far right, with a time of 12:00.

Question 176 B

Paul's rulers cost $60 + 0.2n$, where n is the number of rules made. If he sold n at $1 each, then he received $\$n$.
To break even,

Solutions: B3

$60 + 0.2n = n$
∴ $60 = 0.8n$
∴ $0.8n = 60$
∴ $n = \dfrac{60}{0.8}$
∴ $n = 75$

He must have sold 75 rulers.

Question 177 D

If we add together how much money was charged for the two motorbikes, we get $45. Now, for one motorbike colour we could be charged $15, $20, $25 or $30. For each of the alternatives given to us, here are the expenses;

A. 1 hour plus 2 hours costs $(15 + 20)$
B. 1 hour plus 3 hours costs $(15 + 25)$
C. 1.5 hour plus 2 hours costs $(20 + 20)$
D. 1.5 hour plus 3 hours costs $(20 + 25)$
E. 2 hour plus 3.5 hours costs $(20 + 30)$

Alternative **D** looks good.

Question 178 E

The number of bottles of red wine must be ≥ twice the number of bottles of white wine. That means the expression $x \geq \ldots$ or the expression $\ldots \leq x$ should appear in the answer.

We should also see $2y$ in the answer because we must 'double the number of bottles of white wine'. Only alternative **E** meets these conditions.

Another method might be this; if we have one bottle of white wine, there must be at least two bottles of red wine. Extending this as an example, $x = 1$ and $y = 20$, should satisfy the correct inequation. Only alternative **E** allows such information to be true if you test them all.

Question 179 C

According to the wording of the question, the number 3 we see in the original graph has been cubed which means that $(\sqrt[3]{3}, 1)$ could appear on the answer graph.

Unfortunately, we don't really see that. However, within the graph at the start of the question, the coordinates $\left(1, \dfrac{1}{3}\right)$ could well have appeared. That means on the answer graph, $(\sqrt[3]{1}, \dfrac{1}{3})$ could also appear.

That means $(1, \dfrac{1}{3})$…

Alternative **C** looks good.

Question 180 B

Because two parallel, separate lines don't intersect, look for parallel lines. That is, pairs of straight line equations with equal gradients. One way of checking this is to write down the gradients of each line. For example,
$3x - y = 5$
∴ $y = 3x - 5$
∴ The gradient of this line is 3.
So here is a list of the gradients of the graphs for each alternative.
A. 3 and –4
B. 2 and 2
C. $\dfrac{1}{3}$ and 2
D. $\dfrac{1}{3}$ and $-\dfrac{3}{2}$
E. –4 and 2

From this, it can be see that only alternative **B** provides a pair of parallel lines. For experts, it is not actually good enough that the lines are parallel, they must have different equations.

Solutions: B3

In alternative **B**, the two equations become
$y = 2x - 1$ and
$y = 2x - \frac{3}{2}$. They are separate.

Question 181 A

It is interesting, the information given to us as equations above the graph in the exam is actually not needed at all. All of this information is contained again inside the graph.

Here is a table of the coordinates of each corner of the shaded region. The best (maximum) point will be on a corner.

x	y
25	25
50	50
100	0
50	0

Substituting each pair of numbers into the equation $Z = 2x + ky$ gives the following set of results.

x	y	$Z = 2x + ky$
25	25	50+25k
50	50	100+50k
100	0	200
50	0	100

Now, if the result is that 200 is the biggest value of Z, then neither $50 + 25k$ nor $100 + 50k$ should be as big as 200.

So really we should only be concerned about the value of $100 + 50k$. If k is 2, 3, 4 or 5, it will give the biggest result for Z. The answer, therefore, is that k must be equal to 1.

Solutions: B4

Question 182 C

Sheng pays $120 + 6x = 960$
∴ $6x = 960 - 120$
∴ $6x = 840$
∴ $x = 140$ dollars per month.

Question 183 C

The simple interest formula is:
$$I = P \times \frac{r}{100} \times t$$
Note that 18 months is 1.5 years:
∴ $975 = 26\,000 \times \frac{r}{100} \times 1.5$
∴ $975 = 390r$
∴ $r = 2.5$ per cent per annum.

Question 184 D

$$P' = P(1 + \frac{r}{100})^n$$
$= 45\,000(1 + \frac{4}{100})^5$
$= 54749.38$
The *interest* will be $54749.38 - 45\,000$
$= \$9749.38$
That is nearly $9750 (Keep the change...).

Question 185 C

The annuities formula on your formula sheet is typically programmed into a graphical calculator like this:
$N = 60$
$I\% = 6.2$
$PV = -150\,000$
$PMT = 1100$
$FV = ?$
$P/Y = 12$
$C/Y = 12$
Asking for FV gives an answer of $127\,207.81$ which is nearest to $127\,000$.

Question 186 B

He loses $6500 - 2000 = 4500$ dollars in 5 years. This is $\frac{4500}{5} = 900$ dollars per year depreciation.

Question 187 A

Time 0: $12 200
Time 1:
  $12\,200 + \frac{3}{12}$% of $12\,200 - 5$
$= $12 225.50
Time 2:
  $12\,225.50 + \frac{3}{12}$% of $12\,225.5 - 5$
$= $12 251.06

Question 188 D

Watch that 'quarterly' part: Lim invested $8000 in an investment.
$$\text{Amount} = 8000\left(1 + \frac{r/4}{100}\right)^{5 \times 4}$$
$$= 8000\left(1 + \frac{r}{400}\right)^{20}$$

Question 189 B

Finding how they got that $32.40 is a good start (not in the exam though – you can't afford the time). Watch the word 'minimum' especially for May.
For the April interest:
  $6452.40 \times 0.25/100 = \$16.31$.
For the May and June interest:
  $5992.40 \times 0.25/100 = \$14.98$ twice.
This is a total of $46.09, which is then added to the current total of $5992.40, giving $6038.49.

Question 190 D

The annuities formula is typically programmed into a graphical calculator:

Solutions: B4

$N = ?$
$I\% = 5.6$
$PV = -80\,000$
$PMT = 555$
$FV = 0$ (to pay it all off)
$P/Y = 12$
$C/Y = 12$
This gives an answer of 240 payments or 20 years.
A. Put $N = 120$ and ask for FV. It is $50\,866$ and therefore the loan is not paid.
B. Put $N = 60$ and ask for FV. It is $67\,454$ and not $40\,000$.
C. The amount owing is decreasing so the interest paid each month during the loan is decreasing.
D. At the moment, I calculate N to be 240 to see the loan paid off. If $PMT = 132$ and $P/Y = 52$ and $C/Y = 52$, N must be 983, which is 18.9 years. Thus, weekly repayments of $132 compounding weekly will reduce the period of the loan.
E. Paying an extra payment is always better earlier on. The interest for each subsequent time period will always be smaller as a result.

Question 191 B

For simple interest,
$$I = P \times \frac{r}{100} \times t$$
$$= 37\,000 \times \frac{4}{100} \times 2$$
$$= \$2960$$

Question 192 E

The value of the investment will be
$$P' = P(1 + \frac{r}{100})^n$$
$$= 150\,000(1 + \frac{3.5}{100})^6$$
$$= 150\,000(1.035)^6$$
$$= 184\,388.299$$
$$\approx \$184\,388.30$$

Question 193 E

To increase $36 by 15%, multiply by 1.15. This gives $36 \times 1.15 = 41.4$ per hour. Thus for 2 hours he will charge $82.80 for tutoring.

Question 194 D

The 24 monthly payments total $24 \times 115 = \$2\,760$. Add to this the $500 deposit: the total is $3260.

Question 195 C

$2\,760 is $560 more than the advertised price of $2\,200 (less the deposit). That means an interest rate of
$$I = \frac{P \times r \times T}{100}$$
$$\therefore \quad 560 = \frac{2200 \times r \times 2}{100}$$
$$\therefore \quad 56000 = 4400r$$
$$\therefore \quad r = \frac{56000}{4400}$$
$$\approx 12.7\%$$

Question 196 B

Note that there are 20×12 payments to be made. The annuities formula on your formula sheet is typically programmed into a graphical calculator like this:
$N = 240$
$I\% = 7.4$
$PV = -250\,000$
$PMT = ?$
$FV = 0$
$P/Y = 12$
$C/Y = 12$
Asking for *PMT* gives an answer of $1\,998.72 which is nearest to $1\,999.

Question 197 A

At the end of the second year Binnie's investment is worth

Solutions: B4

$$12\,000\left(1+\frac{3.6}{100}\right)^2 = 12\,879.55.$$

At the end of the second year dear Binnie's investment is worth

$$12\,000\left(1+\frac{3.6}{100}\right)^3 = 13\,343.22.$$

The difference between these results ($463.67) represents the interest earned during the third year. Nearest to answer A.

Question 198 A

After 6 years, the graph must have a height of '5' on its scale. That eliminates C and E. Since the reducing balance method takes off a percentage of the current value each year, the points on the graph should *decrease less* each year. Answer A looks 'just right'.

Question 199 D

Watch out for the word 'not'...
Note that there are 10×12 payments to be made. The annuities formula on your formula sheet is typically programmed into a graphical calculator like this:
$N = 120$
$I\% = 7.5$
$PV = -130\,000$
$PMT = ?$
$FV = ?$
$P/Y = 12$
$C/Y = 12$

In part A, 7.5% divided by 12 is indeed 0.625%. This is true.
In part B, if we are told that the loan is fully repaid after 10 years, then yes, I would expect nothing is owing after 10 years. True.
In part C, yes, there are 120 payments. True.
In part D, if you put 1500 into *PMT* as shown above and then solve for *FV*, you will see $7672.89. That means that the loan is not paid off at 10 years. D is false

In part E, start by solving for *PMT* so the future value *FV* will be zero. This gives $1543.122 as the payment. Keeping this value, now change *N* to 60 (five years) and solve for *FV*. This gives $77 010.03, which certainly exceeds $65 000. True.

Question 200 D

Simple interest:
$$I = \frac{P \times r \times T}{100}$$
$$\therefore \quad 140.38 = \frac{1200 \times r \times 1}{100}$$
$$\therefore \quad 14038 = 1200r$$
$$\therefore \quad r = \frac{14038}{1200}$$
$$= 11.699$$
$$\approx 11.70\%$$

Question 201 B

$$\text{New price} = \text{old price} \times \left(1 + \frac{10}{100}\right)$$
$$\therefore \quad 22 = x \times 1.1$$
$$\therefore \quad x = \frac{22}{1.1} = \$20$$

Question 202 B

The second balance is 2143.50 + 2.45 which is 2145.95. The second balance will be 2145.95 − 616.40 = 1529.55. She then jumped to 1971.75 so that her salary was 1971.75 − 1529.55 = 442.20.

Question 203 A

The depreciation will be 240 000 units multiplied by $0.05 giving $12 000. This reduces the value to
45 000 − 12 000 which is $33 000.

Question 204 B

This is the perfect example for TVM solver on your graphical calculator:
$N = 60$

Solutions: B4

$I\% = 8.5$
$PV = -200\,000$
$PMT = ?$
$FV = 200\,000$
$P/Y = 12$
$C/Y = 12$
Solving gives the PMT monthly payment as $1416.66666 or, more sensibly, $1416.67.

Question 205 C

Apply this formula:
New principal = old principal $\times 1.065^n$
where n is the year number.
Here is a list of the year number along with the value of his money:

Year	Amount
0	$3000.00
1	$3195.00
2	$3402.68
3	$3623.85
4	$3859.40

We are looking for the increase from year 3 to year 4. That is, $3623.85 – $3859.40 which is $235.55.

Question 206 A

Again, it is best to use TVM solver here:
$N = 16$
$I\% = 6$
$PV = -10\,000$
$PMT = 500$
$FV = ?$
$P/Y = 12$
$C/Y = 4$
Solving gives the FV future value as $3723.670555 or 'closest to $3720'.

Question 207 D

It is smart to start on Monday with $1. The value of that dollar on each of the days is found by multiplying the previous day's value in turn by 1.10, 0.90, 1.20, 0.80. The results are:

Monday	1
Tuesday	1.1
Wednesday	0.99
Thursday	1.188
Friday	0.9504

The dollar has dropped by about 5%.

Question 208 E

This question can be done without a calculator and without information about the interest rate of amount borrowed. In the missing month of payment, the amount owed will not have been reduced. Therefore the amount of interest paid will be calculated on a higher amount. This means the loan will not be paid off as it would have been done if the $800 had been paid properly.

Question 209 B

Simple interest = $P \times r \times t$
$= 4000 \times \dfrac{5}{100} \times 1$
$= \$200$

Question 210 C

The data is for the month of October (only). The minimum balance of the month is $473.92. The interest on this is 0.15%
$\therefore\quad 473.92 \times \dfrac{0.15}{100}$
$= 0.71088$
$\approx \$0.71$

Question 211 E

We are told that the simple interest Grandad earned on his $P invested is $584 per month. The interest rate of 6.2% must be divided by 12 ('monthly').
Simple interest = $P \times r \times t$
$\therefore\quad 584 = P \times \dfrac{6.2}{100 \times 12} \times 1$
$\therefore\quad 584 = P \times 0.00516666$

Solutions: B4

$$\therefore \quad P = \frac{584}{0.00156666}$$
$$= \quad 11\,303.2404$$
Nearest answer: $113 000.

Question 212 B

To increase an amount of money by 10%, multiply it by 1.10.
$$\text{New} = \text{old} \times 1.1$$
$$\therefore \quad 825 = \text{old} \times 1.10$$
$$\therefore \quad \text{old} = \frac{825}{1.10}$$
$$= \quad 750.$$
That means GST of $825 - 750 = 75 was added to the original amount.

Question 213 E

The value of the depreciation was
$$48\,000 - 21\,000$$
$$= \quad \$27\,000$$
But depreciation $= 0.04 \times$ copies
$$\therefore \quad \text{copies} = \frac{\text{depreciation}}{0.04}$$
$$= \quad \frac{27\,000}{0.04}$$
$$= \quad 675\,000 \text{ copies}$$

Question 214 C

The person paid $200 plus 36×68
$$= \quad \$2648$$
This is $648 bigger than the $2000 price. Now, the amount borrowed is
$2000 - 200 = 1800. The unit of time is the year, so the loan was for 3 years.

$$\text{Simple interest} = P \times r \times t$$
$$\therefore \quad 648 = \frac{1800 \times r \times 3}{100}$$
$$\therefore \quad 648 = 54r$$
$$\therefore \quad r = 12\%$$

Question 215 D

Call the original price P. The 20% discount means that the sale price was 80% of the original or $0.80P$. Now subtract the $80 trade-in. This gives $368.
$$\therefore \quad 0.80P - 80 = 368$$
$$\therefore \quad 0.80P = 448$$
$$\therefore \quad P = \frac{448}{0.8}$$
$$= \quad \$560.00$$

Question 216 A

In each year, there are four equal bank balances. That means that any interest is paid at the end of the year ('credited annually'). Note how the dots rise up by 'one square' after 1 year but rise up by more than one square in the next year and so on. That means compound interest is operating, not simple interest. The first alternative is correct.

Question 217 B

For this question, it is best to use the annuities formula in the graphical calculator. First, we will need to see how much the monthly payments are. It is paid off after 60 payments. The values you will need are:
N = 60
I% = 9.2
PV = –18 000
PMT = ?
FV = 0
P/Y = 12
Asking the calculator for the monthly payment produces an answer of $375.40. Now we need to repeat the process for an N value of 10:
N = 10
I% = 9.2
PV = –18 000
PMT = 375.4
FV = ?
P/Y = 12

Solutions: B4

Asking the calculator for the future vale (FV), produces an answer of $15 542.40 which means that Jenny has paid off $18\,000 - 15\,542.40 = \$2457.60$.

Question 218 A

The selling price is $500 + 20 = \$520$. The required percentage is

$$\frac{20}{500} \times \frac{100}{1}$$
$$= 4\%$$

Question 219 C

If the value of the car to drops to $8000, then $22 000 has been lost. That means

$0.25 \times$ kilometres $= 22\,000$

$\therefore$ kilometres $= \dfrac{22\,000}{0.25}$

$= 88\,000$ km.

Question 220 C

Simple interest $= \dfrac{P \times r \times t}{100}$

$\therefore \ 27\,000 = \dfrac{P \times 8 \times 6}{100}$

$\therefore \ \dfrac{P \times 48}{100} = 27\,000$

$\therefore \ P = \dfrac{27\,000 \times 100}{48}$

$= \$56\,250$

Question 221 D

We are interested in the third line of the table. The stamp duty on $200 000 is $2560 plus 6% of the price in excess of $115 000.
That is, 6% of $200\,000 - 115\,000$.

$= \dfrac{6}{100} \times 85\,000$

Add the $2650 to equal $7660.

Question 222 E

For this question, I will look at each alternative separately.

A. The kitchen depreciates by $4000 annually? No, the curve ensures that it drops by a separate amount every year.
B. No, at the end of five years, the value is about $22 500.
C. At the start, the value dropped by about $7500. Compared to $50 000, that is about 15%. So no, it is not always below 5%.
D. The annual depreciation rate increases over time.
E. Because the curved graph is flattening out, the amount of money lost each year *is* decreasing.

Question 223 D

The compound interest formula states that $A = P \times R^n$ where, in this case, $R = 1 + \dfrac{0.5 \times 10}{100} = 1.05$, noting that because the time period is six months the number of times the money compounds is 10.
Thus, after five years, the principal will become $10\,000 \times 1.05^{10}$. It was so tempting to get out my calculator at this stage…

Question 224 E

The original unknown amount, x, would have been multiplied by 1.03 at the end of the first year, and the result multiplied by 1.02 at the end of the second year. The result is $42 000.

$\therefore \ 42\,000 = x \times 1.03 \times 1.02$

$\therefore \ x = \dfrac{42000}{1.03 \times 1.02}$

$= \$39\,977$.

Note: be careful with the use of brackets (or of the division) when calculating $\dfrac{42000}{1.03 \times 1.02}$.

Question 225 D

If Brad buys the $720 washing machine, his deposit of $180 means that he will only be contracted for $540 under the hire purchase agreement.
The flat rate of 12% means that
$$2 \times \frac{12}{100} \times 540 = \$129.6$$
must be added to his costs. So, the monthly payment will be $\frac{540 + 129.6}{24} = \$27.90.$

Question 226 A

The annuities formula on your formula sheet is typically programmed into a graphical calculator like this:
$N = 240$
$I\% = 7$
$PV = -250\ 000$
$PMT = ?$
$FV = 0$
$P/Y = 12$
$C/Y = 12$
Asking for *PMT* gives a monthly payment value of $1938.25. multiply this by 240 and you will find that Petra will pay $465 179.36. The interest she has paid will be
$$= \begin{array}{l} \$465\ 179.36 - 250\ 000 \\ \$215\ 179.36 \\ \text{or about } \$215\ 000. \end{array}$$

Solutions: B5

Question 227 C

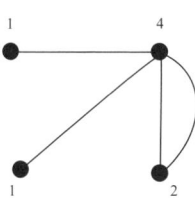

The degree of the four points are shown here.
Their sum is 8.

Question 228 D

An Euler **path** is where you can move along all paths, but along each path only once. By removing the edge from *B* to *C*, the order of all vertices becomes even (as required). The path I found most easily is the path shown here:

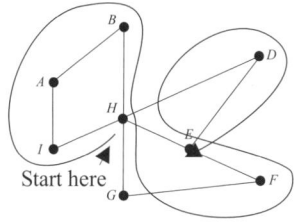

Question 229 B

Look for differences in the matrices. Kael to *V* is not possible. There goes **A**. Lian to *X* is possible but not to *V*. Reject **C** and **E**. Minh to *X*? No. That only leaves **B**.

Question 230 B

Using the 5 above the tank is sensible. That leads immediately to the 6, 8 and 9. Underneath the tank, using the 7 is best. The 4, 4, 9 and 8 are the best there. It may be useful to check that the other answers are worse.

Question 231 A

Euler's formula says that $V = E - F + 2$.
Start with $V = 10$ and $E = 15$.
Now in the altered state, $V = 10$ and $F = 3$.
$$10 = E - 3 + 2$$
$$\therefore \quad 10 + 3 - 2 = E$$
$$\therefore \quad E = 11$$
This is a reduction of $15 - 11 = 4$ edges.

Question 232 C

Cut A has flow equal to $7 + 5 + 5 + 3 + 2 = 22$. Cut B has flow $7 + 5 + 5 + 0 + 2 + 2 = 21$. Cut C has flow $7 + 5 + 6 = 18$, cut D is not useful as it misses the lower node and Cut E has flow $18 + 6 = 24$. The *smallest* one is 18 for cut C.

Question 233 D

A is rejected because there are 3 roads from *M* to *L*.
B is rejected because there are 3 roads from *M* to *L*.
C is rejected because there is no direct loop at M.
E is rejected because there is no direct loop at M.
That leaves **D** which actually looks good...

Question 234 C

The critical path is *R-U-X-Z*. If any of those path lengths then is increased, the time will increase. Hence answer C.

Question 235 B

It may help to put all of the data in the matrix on a diagram:

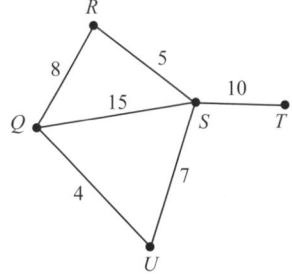

Here are the possible paths and their lengths:

193

Solutions: B5

Q-S-T has length 15 + 10 = 25
Q-R-S-T has length 8 + 5 + 10 = 23
Q-U-S-T has length 4 + 7 + 10 = 21
That is, route Q-U-S-T is shortest with length 21 kilometres.

Question 236 B

The order of each vertex is noted here:

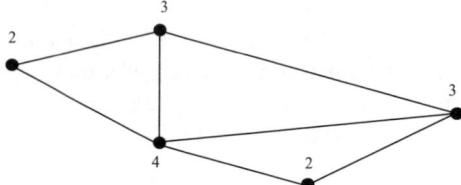

Two of the vertices have an odd order.

Question 237 B

It is best to try each alternative.
Starting at intersection 4, you can't get to intersection 5.
Starting at intersection 5, you *can* get to them all.
Starting at intersection 6, you can't get to intersection 5.
Starting at intersection 7, you can't get to intersection 5.
Starting at intersection 8, you can't get to intersection 5.

Question 238 D

The first alternative doesn't end up where it started and also, I found some Hamiltonian paths. Alternative C misses the edge GE.
Exactly two vertices have an odd order (B and G). Therefore Eulerian paths exist. I found *BEFGEDCBAG* and *GFEBCDEGCAG*.

Question 239 D

This is the minimal spanning tree that I found:

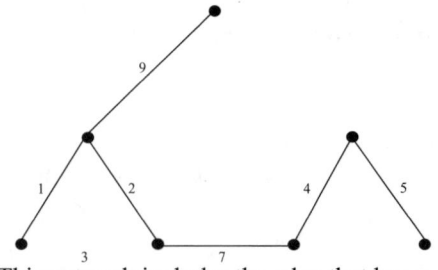

This network includes the edge that has a weight of 9 and none of the edges that have weights of 3, 6, 8 or 10.

Question 240 E

Look for a fault in each alternative.
A.   Imm. predecessor of *I* is not *G*.
B.   Imm. predecessor of *J* is not *H*.
C.   E is also an imm. predecessor of *I*.
D.   Imm. predecessor of *I* is not *G*.
E.   Good answer!

Question 241 C

All of the edges pass from the source side to the sink side except for *d*. It does not contribute to the flow. Alternative C demonstrates this.

Question 242 C

In a complete graph, each vertex must be joined to each other vertex. In A and B, the vertices have only 2 (or 1) edges. It is possible D is correct but it is not the best answer. E has no edge joining students. C is just right.

Question 243 D

Let v = vertices, f = faces and e = edges, Euler ($v + f = e + 2$) applied to each gives:

A. $4 + f = 5 + 2$
B. $6 + 5 = 9 + 2$
C. $6 + 7 = 11 + 2$
D. $5 + 4 \neq 10 + 2$ (not planar)
E. $7 + 5 = 9 + 2$

Solutions: B5

Question 244 B

First calculate the various paths from the start to the finish:
$B + G + I + J = 5 + 3 + 8 + 1 = 17$
$A + F + J = 1 + 12 + 1 = 14$
$A + C + E + H + J = 1 + 2 + 7 + 4 + 1 = 15$
The earliest completion time is 17 hours. The best path to crash is *BGI*. It will reduce the time to match the next early completion time of 15 hours, That is, a reduction of 2 hours.

Question 245 E

I will comment on each alternative. Please note the special word 'not' in this question.

A. Complete. Yes, there are no isolated nodes.
B. Planar. Yes, if you separate one of the 'diagonal lines' and make it go around the outside, then this graph can be drawn on a plane surface without any crossovers.
C. Simple. Yes, it is.
D. Undirected. True, there are no arrows.
E. Tree. No, it is not a tree. Trees do not have loops. This diagram has loops.

Question 246 C

Euler's formula can be applied here.
$$v + f = e + 2$$
But since there are 12 edges,
$\therefore \quad v + f = 14$
So now look at the alternatives and find one where v and f add up to 14. **C** is the only alternative that does this.

Question 247 A

The structure of an adjacency matrix is like this:

$$\begin{array}{c} \text{To} \\ \text{From } A \\ B \\ C \\ D \end{array} \begin{array}{cccc} A & B & C & D \\ \left[\begin{array}{cccc} ? & ? & ? & ? \\ ? & ? & ? & ? \\ ? & ? & ? & ? \\ ? & ? & ? & ? \end{array}\right] \end{array}$$

You would expect symmetry left and right of the main diagonal. A careful look shows that only alternative **A** survives this test. Go no further…

Question 248 B

The shortest distance connecting town *A* to town *B* that I found was along this path:

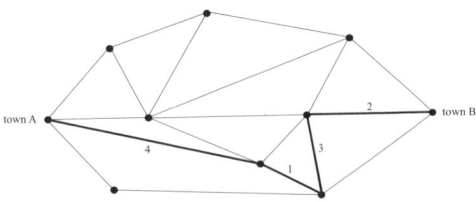

The total distance is $4 + 1 + 3 + 2 = 10$ km.

Question 249 D

A comment on each alternative is appropriate here.

A. Activities *A*, *B* and *C* all have the same earliest start time. Yes, this is true. They all start at time zero.
B. There is only one critical path for this project. Yes, it is path *C, D, F, K, L*.
C. Activity *J* may finish at any time from 11 to 16. That means it may start up to 5 hours later than the 5 hours shown.
D. No, the critical path determines the minimum time for project completion.
E. Activity *L* must be on the critical path because as the last 'leg' of the path, it can't be avoided.

Solutions: B5

Question 250 E

'Activity' D must be included. Thus, the critical path C, D, F, K, L. determines the times. It arrives at the start of activity L at time 16 hours.

Question 251 C

This question can be quite time consuming. You could write down the diagram 5 times with the values of x and y taken from each alternative written in the appropriate place in each diagram. You would then find the length of the minimal spanning tree and see whether it is equal to 19. Here is the diagram for alternative **C**.

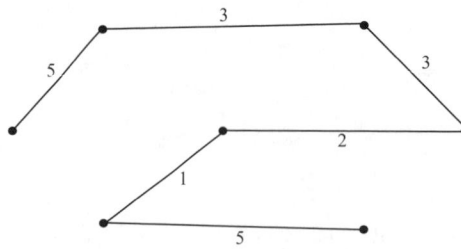

Check that it adds to 19. Note that '$x + y = 8$' is only true in alternative C.

Question 252 B

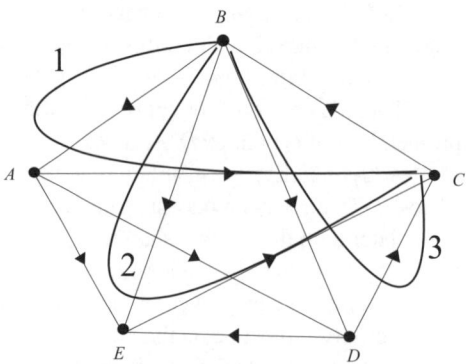

For a two-step dominance question, it might be useful to set out a matrix like this;

$$\text{From} \begin{array}{c} \\ A \\ B \\ C \\ D \\ E \end{array} \begin{array}{c} \text{To} \quad A \; B \; C \; D \; E \\ \left[\begin{array}{ccccc} & & & & \\ & & & & \\ & & & & \\ & & & & \\ & & & & \end{array} \right] \end{array}$$

From B to C, there are three two step paths. The full matrix showing all of the two-step paths follows. Note that the 'B- row' has the largest number of two step dominance.

$$\text{From} \begin{array}{c} \\ A \\ B \\ C \\ D \\ E \end{array} \begin{array}{c} \text{To} \quad A \; B \; C \; D \; E \\ \left[\begin{array}{ccccc} 0 & 0 & 2 & 0 & 1 \\ 0 & 0 & 3 & 0 & 2 \\ 0 & 0 & 0 & 1 & 1 \\ 0 & 1 & 1 & 0 & 0 \\ 0 & 1 & 0 & 0 & 0 \end{array} \right] \end{array}$$

Question 253 B

For 2007, consider team A. They had a choice of four teams to play. Having noted those choices, team B had only three choices of competitors, because the game of A versus B had already been counted. Team C would have had two choices and team D would only have had one choice. Thus, for 2007, there were
$4 + 3 + 2 + 1 = 10$ games.
In 2008, this will increase. This time, there will be $6 + 5 + 4 + 3 + 2 + 1 = 21$ choices of game. That is an increase of 11 games compared to 2007.

Solutions: B6

Question 254 B

$$\begin{bmatrix} 12 & 36 \\ 0 & 24 \end{bmatrix}$$

Take out a common factor of 12:

$$= 12\begin{bmatrix} 1 & 3 \\ 0 & 2 \end{bmatrix}$$

Question 255 D

The order or the matrices are:
A: 2×1
B: 1×2
C: 1×1
Now consider each alternative:

A. AB 2×1 and 1×2. Defined.
B. AC 2×1 and 1×1. Defined.
C. BA 1×2 and 2×1. Defined.
D. BC 1×2 and 1×1. Not defined.
E. CB 1×1 and 1×2. Defined.

Question 256 A

The matrix A is the unit matrix. So $A^3 = I$.

$$B - C = \begin{bmatrix} 2 & 1 \\ 1 & 0 \end{bmatrix} - \begin{bmatrix} 1 & -1 \\ -1 & 1 \end{bmatrix}$$

$$= \begin{bmatrix} 1 & 2 \\ 2 & -1 \end{bmatrix}$$

That means that $A^3(B-C)$ is $\begin{bmatrix} 1 & 2 \\ 2 & -1 \end{bmatrix}$.

Question 257 B

Reading across the top line of that first table, in 2004, Blue won 1, Green won 2 and Red won 0.
In 2005, Blue won 1, Green won 1 and Red won 1.
In 2006, Blue won 2, Green won 1 and Red won 0.

Thus, the matrix $\begin{array}{c} \\ 2004 \\ 2005 \\ 2006 \end{array}\begin{array}{ccc} B & G & R \\ \end{array}$
$\begin{bmatrix} 1 & 2 & 0 \\ 1 & 1 & 1 \\ 2 & 1 & 0 \end{bmatrix}$

is correct.

Question 258 D

We require the matrix product MP. Note that the product PM is not possible.

$$MP = \begin{bmatrix} 1.2 & 0 \\ 0 & 1.35 \end{bmatrix} \times \begin{bmatrix} 145 & 210 & 350 \\ 185 & 270 & 410 \end{bmatrix}$$

$$= \begin{bmatrix} 174 & 252 & 420 \\ 249.75 & 364.5 & 553.5 \end{bmatrix}$$

Question 259 E

Write the matrix orders under the expression

$$XA = \begin{bmatrix} 4 & 1 \\ 1 & 4 \\ 3 & 5 \end{bmatrix}:$$

$(?\times ?) \times (3\times 2) = (3\times 2)$
Both '?' values must be the number 3.

Question 260 C

To have a unique solution, two lines must *not* be parallel. Here are the 5 pairs of equations again, each with the gradient written next to it:

equation	gradient
$4x + 2y = 10$	-2
$2x + y = 5$	-2
$x = 0$	0
$x + y = 6$	-1
$x - y = 3$	1
$x + y = 3$	-1
$2x + y = 5$	-2
$2x + y = 10$	-2

Solutions: B6

$x = 8$ ∞
$y = 2$ 0

Of these pairs, 2 are parallel and 3 are not. Those are the 3 pairs we want.

Question 261 E

Look for a matrix whose structure is

$$\begin{array}{c c} & \text{Australia Overseas} \\ \text{Australia} & \begin{bmatrix} ? & ? \\ ? & ? \end{bmatrix} \\ \text{Overseas} & \end{array}$$

The cells must add vertically to give 1 and the 0.95 and 0.20 must be diagonally opposite. Only E meets these requirements.

Question 262 B

This transition matrix should be raised to a power. I entered it as a matrix in my graphical calculator and raised it to the power of 30. It gave $\begin{bmatrix} 0.001 & 0 \\ 0.999 & 1 \end{bmatrix}$. The birds in location A are disappearing. Gradually.

Question 263 A

Add these by adding the corresponding elements.

$$\begin{bmatrix} 0 & -4 \\ 2 & 5 \end{bmatrix} + \begin{bmatrix} 5 & 4 \\ -2 & 2 \end{bmatrix}$$

$$= \begin{bmatrix} 0+5 & -4+4 \\ 2-2 & 5+2 \end{bmatrix}$$

$$= \begin{bmatrix} 5 & 0 \\ 0 & 7 \end{bmatrix}$$

Question 264 D

A 3×1 matrix looks like this: $\begin{bmatrix} ? \\ ? \\ ? \end{bmatrix}$. That eliminates alternatives **A**, **B** and **E**. The data must all come from 2005, so that eliminates **C**. Alternative **D** is 'just right'.

Question 265 B

There are probably quicker ways, but I will set this question out formally, using algebra.

$$AX = \begin{bmatrix} 5 & 6 \\ 8 & 10 \end{bmatrix}$$

$$\therefore \begin{bmatrix} 8 & 4 \\ 5 & 3 \end{bmatrix} X = \begin{bmatrix} 5 & 6 \\ 8 & 10 \end{bmatrix}$$

$$\therefore X = \begin{bmatrix} 8 & 4 \\ 5 & 3 \end{bmatrix}^{-1} \begin{bmatrix} 5 & 6 \\ 8 & 10 \end{bmatrix}$$

$$= \frac{1}{24-20} \begin{bmatrix} 3 & -4 \\ -5 & 8 \end{bmatrix} \begin{bmatrix} 5 & 6 \\ 8 & 10 \end{bmatrix}$$

$$= \frac{1}{4} \begin{bmatrix} 15-32 & 18-40 \\ -25+64 & -30+80 \end{bmatrix}$$

$$= \frac{1}{4} \begin{bmatrix} 17 & -22 \\ 39 & 50 \end{bmatrix}$$

Note the decimals in the answers…

$$= \begin{bmatrix} -4.25 & -5.5 \\ 9.75 & 12.5 \end{bmatrix}$$

Other methods include using a calculator.

Question 266 B

The equations:
$2x + z = 5$
$x - 2y = 0$
$y - z = -1$

It is not a bad idea to re-write them:
$2x + 0y + z = 5$
$x - 2y + 0z = 0$
$0x + y - z = -1$

In matrix form:
$$\begin{bmatrix} 2 & 0 & 1 \\ 1 & -2 & 0 \\ 0 & 1 & -1 \end{bmatrix} \begin{bmatrix} x \\ y \\ z \end{bmatrix} = \begin{bmatrix} 5 \\ 0 \\ -1 \end{bmatrix}$$

Question 267 A

It is probably best to analyse the 5 alternatives one by one. The correct answer must be a single number (the total value of silver). In addition, the matrix pairs should conform to the rules of matrix

198

multiplication. Here are the orders of the matrix pairs and the order of each corresponding result.

A. $(1\times 3)\times(3\times 1)$ gives (1×1)
B. $(1\times 3)\times(1\times 3)$ **not possible **.
C. $(3\times 1)\times(3\times 1)$ **not possible **
D. $(3\times 1)\times(1\times 3)$ gives (3×3)
E. $(1\times 3)\times(3\times 1)$ gives (1×1)

So, alternatives **A** and **E** both seem possible. However, in **E** we appear to be multiplying the intermediate gold, silver and bronze award values by the number of silver prizes. This is hardly logical. Alternative **A** precisely multiplies the silver prize values by the number of corresponding winners. The answer is **A**. For those desperate to know the total money value, it is $2240. Calculating that (unwanted) answer was a definite waste of my time, but I enjoyed it…

Question 268 D

Consider the matrix $T = \begin{bmatrix} 0.8 & 0.1 & 0.2 \\ 0.1 & 0.6 & 0.1 \\ 0.1 & 0.3 & 0.7 \end{bmatrix}$.

Define the matrix Z to be $\begin{bmatrix} 1568 \\ 1105 \\ 894 \end{bmatrix}$.

The result of multiplying $T \times Z$ will be a matrix defining the number of bats in each location on the second night. Similarly, the result of multiplying $T \times T \times Z = (T^2 Z)$ will be a matrix defining the number of bats in each location on the third night.
If this process is repeated 20 times and then is repeated one more time, the corresponding matrices will be

$T^{20}Z = \begin{bmatrix} 1605.13 \\ 713.40 \\ 1248.46 \end{bmatrix}$

and

$T^{21}Z = \begin{bmatrix} 1605.14 \\ 713.40 \\ 1248.46 \end{bmatrix}$.

That is, the multiplication result has reached a stable answer by the 20th (night). The first value in that answer matrix represents the number of bats feeding at location A. It is 1605 bats.

Question 269 E

To qualify as a transition matrix, the elements in each vertical column must add to 1. That eliminates alternative **D**. for this problem, we are told that the family does not visit the same port two years running. So, for example, Portland this year and Portland next year is not acceptable. The elements down the main diagonal (the numbers form the top left to the bottom right) must all be zero. Alternative **E** meets these conditions,

Question 270 B

We are given the transition matrix that must multiply the matrix showing the previous answer that was chosen. The first such matrix is $\begin{bmatrix} 0 \\ 0 \\ 0 \\ 1 \end{bmatrix}$ because we are told that Kerry's alternative **D** was the first answer. For the second answer, multiply the transition matrix by this matrix:

$\begin{bmatrix} 1 & 0 & 1 & 0 \\ 0 & 0 & 0 & 1 \\ 0 & 1 & 0 & 0 \\ 0 & 0 & 0 & 0 \end{bmatrix} \begin{bmatrix} 0 \\ 0 \\ 0 \\ 1 \end{bmatrix} = \begin{bmatrix} 0 \\ 1 \\ 0 \\ 0 \end{bmatrix}$

You may interpret this matrix by stating that alternative **B** was chosen next. Next,

Solutions: B6

$$\begin{bmatrix} 1 & 0 & 1 & 0 \\ 0 & 0 & 0 & 1 \\ 0 & 1 & 0 & 0 \\ 0 & 0 & 0 & 0 \end{bmatrix} \begin{bmatrix} 0 \\ 1 \\ 0 \\ 0 \end{bmatrix} = \begin{bmatrix} 0 \\ 0 \\ 1 \\ 0 \end{bmatrix}$$

That is, alternative **C** was next.
Then,

$$\begin{bmatrix} 1 & 0 & 1 & 0 \\ 0 & 0 & 0 & 1 \\ 0 & 1 & 0 & 0 \\ 0 & 0 & 0 & 0 \end{bmatrix} \begin{bmatrix} 0 \\ 0 \\ 1 \\ 0 \end{bmatrix} = \begin{bmatrix} 1 \\ 0 \\ 0 \\ 0 \end{bmatrix}$$

That is, alternative **A** was next.
Then,

$$\begin{bmatrix} 1 & 0 & 1 & 0 \\ 0 & 0 & 0 & 1 \\ 0 & 1 & 0 & 0 \\ 0 & 0 & 0 & 0 \end{bmatrix} \begin{bmatrix} 1 \\ 0 \\ 0 \\ 0 \end{bmatrix} = \begin{bmatrix} 1 \\ 0 \\ 0 \\ 0 \end{bmatrix}$$

That is, alternative **A** was next.
So far we have that alternatives **D**, **B**, **C**, and **A** have been chosen which is all that is needed. Always keep an eye on the answers. So far, we have uniquely matched *our* exam paper's alternative **B**. We can stop.

Hmm. Quite an interesting way to answer an exam paper, don't you think? Simply take a transition matrix into the exam with you. Oh, but you must know what the first answer is…

Question 271 C

This question is asking you to identify the rules for matrix multiplication.
We are told that $M \times N \times P$ is a feasible multiplication. Note that the brackets do not matter for multiplication.
If $M \times N \times P$ is feasible, then the orders must be $(3 \times 4) \times (a \times b) \times (5 \times c)$, where a, b and c are three pronumerals that I have made up. The other information I used to make that statement is supplied in the question.
Immediately, we can tell that $a = 4$ and $b = 5$. Now, the question asked for the order of N. Using my definition, it is $(a \times b)$ or 4×5.
Note that the value of pronumeral c is not needed.

Solutions: C1

Question 272

a. When you plot the required points, the result looks like this:

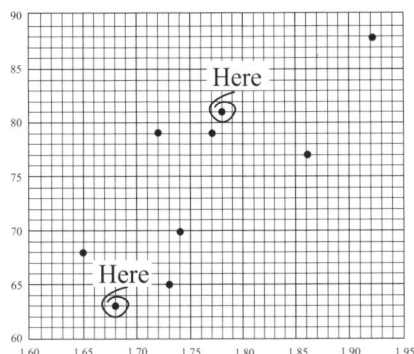

b. On my calculator, I entered the height data into the first column and the weight data into the second column. Fitting a straight line gave the result of $y = -60.821 + 76.807x$. Those numbers should be corrected to 1 decimal place and the correct variables should be used: $weight = -60.8 + 76.8 \times height$.

c. For this sample, 61% of the variation in the weight of the people can be explained by the variation in their height.

Question 273

a. $BMI = \dfrac{weight}{(height)^2}$
$= \dfrac{66}{1.69^2}$
$= 23.10843$
≈ 23.1

b. Both columns of *BMI* data are sorted. For **males**, the lowest *BMI* is 20.6 and the highest is 31.4. Thus, the range is $31.4 - 20.6 = 10.8$.

c. The overweight people are above 25.0. Counting them gives: Males 6, females 3. Since there are 17 males, 11 must be 'not overweight'. Similarly, there must be 18 who are 'not overweight'.

Filling in the table,

	male	female
overweight	6	3
not overweight	11	18

d. We will convert this to percentages. Divide the 'male numbers' by 17 and the 'female numbers' by 21, and then multiply by 100:

	male	female
overweight	35%	14%
not overweight	65%	86%

Note that 'raw numbers' alone cannot be used here. The results support the idea that there are gender issues with respect to 'BMI overweight' measurements. Men seem a little heavier. Of course...

e. i. The medians are very similar. The *IQR* values are reasonably similar. Not much else...

ii. I entered the data in the first column of my 'calculator spreadsheet' and used 1-variable statistics to find $\bar{x}$. The answer, rounded to 1 decimal place (don't forget this), was 23.9.

iii. It is reasonable to assume that there are outliers in the male data. That would explain the skew (asymmetry) in the male boxplot. The statistic median is not affected by extreme values and thus would be better than mean here. Mean *is* affected by outliers. Median is best here.

Question 274

a. It seems logical that the value of a car depends on how old it is. Thus *value* is the dependent variable.

Solutions: C1

b. i. The coefficient of determination is the square of the correlation coefficient. Thus, we must find the square root of 0.9058 which is ±0.9517, or ±0.952 correct to three decimal places. But clearly, there is a negative trend. Hence −0.952.

ii. The percentage is
$$0.9058 \times \frac{100}{1} = 90.58\% \approx 91\%.$$

c. Let the equation of the least squares regression line be $y = a + bx$. The y-intercept of the graph shown is 17 500 and the (negative) gradient can be found by fitting a '$\frac{\text{rise}}{\text{run}}$' triangle to the diagram:

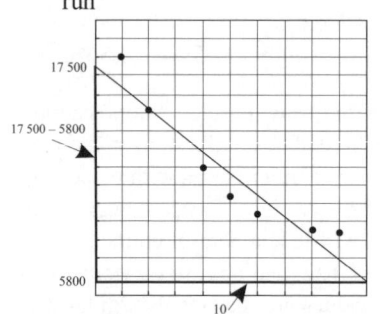

You must estimate those y-numbers (value in dollars). The gradient is
$$-\frac{17500 - 5800}{10} = -1170.$$ The equation, with coefficients correct to the nearest hundred, becomes
value = 17 500 − 1200 × age

d. Residual = observed − predicted. For the predicted vales, use the equation $y = 17500 - 1170x$ with $x = 4$, 5 and 6 years. I got the following answers: 12 820, 11 650 and 10 480. Now make a table for these:

	observed	predicted	residual
$x = 4$	11 900	12 700	−800
$x = 5$	10 400	11 500	−1100
$x = 6$	9600	10 300	−700

Adding these values to the residual plot gives the following:

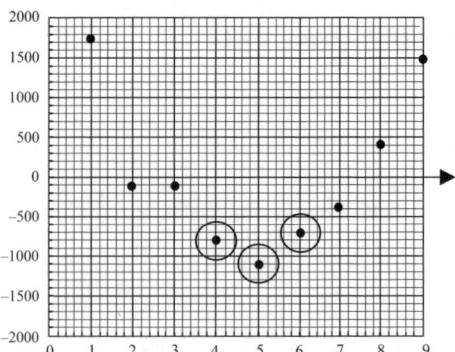

The new 'dots' have been circled here for clarity. Note that the same process might have been completed by estimating from the graph given in the exam how far the dots are under the line. It would have been a less accurate method, but much faster.

e. The pattern of dots on the residual form a curved graph, perhaps a parabola. Definitely, not a linear relationship.

f. The missing number is
$\log_{10}(9) = 0.954243$
≈ 0.95

g. The new graph shows a strong negative linear relationship between value and log(age). The trend is a decreasing one.

h. Use the equation given with $age = 3$.
value = 18 300 − 10 800 × log(age)
= 18 300 − 10 800 × log(3)
= 18 300 − 5152.91
= 13147.09
≈ $13 100 (to the nearest hundred dollars).

i. This a random pattern of dots. The curved pattern has been removed.

j. A more complex answer: the opposite operation to taking $\log_{10}()$ of the age values is to raise 10 to the power of the value axis. The numbers involved would be quite big…

Solutions: C1

A more simple answer: The dots need to be squashed towards the vertical ('y') axis to make the dots form more of a straight line. Or you could also say squash the dots towards the horizontal 'x-axis'.

Question 275

a. I typed the '18 year-old' numbers in the 'spreadsheet' section of my graphical calculator and pressed Stat, Calc. 1-Var Stats. and found that $\bar{x}$ was the same as theirs (82.67 rounded to 82.7). The correct version of the standard deviation is found against $s_x = 3.84155 \approx 3.8$. The answer is *not* 3.7.

b. Be sure to use the 27 month data. To convert an X score to a Z score, use

$$z = \frac{x - \mu}{\sigma}$$
$$= \frac{83.1 - 89.3}{4.5}$$
$$= -1.3777$$
$$\approx -1.4.$$

Also be sure to round to 1 decimal place.

c. I will express this as a diagram:

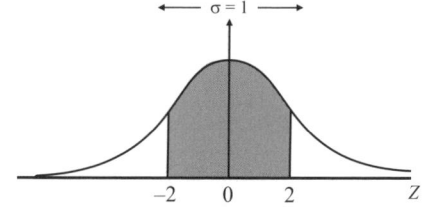

The area between -2 and 2 on such a diagram is 95% of the total area. That means the white area is half of 5% which is 2.5%. The shaded area is $100 - 2.5 = 97.5\%$.

d. For the 18-month-old boys, the required statistics are:
Min = 76 cm
Max = 89.8 cm
Median $= \dfrac{82.8 + 83.2}{2} = 83$ cm
$Q_1 = 80$ cm

$Q_3 = 85.8$ cm
You may wish to check those values using the method you have been taught. The resulting boxplot is added to the group:

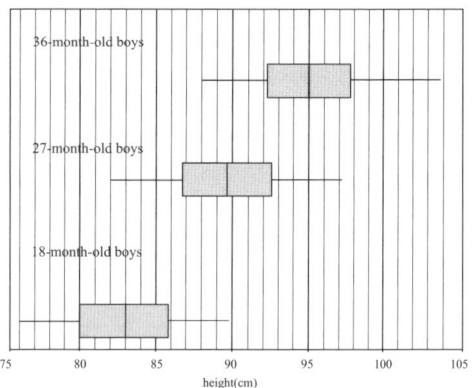

e. From the middle boxplot, the median (the vertical line inside the shaded box) is at about 89.5 cm.

f. The most appropriate statistic is the median – and it asks for *an* appropriate statistic. So my answer is that the median height increases with age. However, the fact that all five numbers that are needed to construct the boxplot increase with age might also be a suitable fact. It increases with age. Another important idea is the shape of the boxplots remaining the same with age.

Question 276

a. The gradient of the line is 0.53. That means that the height increases by 0.53 cm for each month of age increase. The missing number is 0.53.

b. **i.** The coefficient of determination is the square of the correlation coefficient multiplied by 100.
$$0.7541^2 \times 100$$
$$= 56.866681$$
$$\approx 56.9\%$$

Solutions: C1

ii. 56.9% of the variation of height is *explained* by the variation in age. Note that it is not definitely *caused* by that variation.

Question 277

a. I hope you will pardon all of my construction lines below. I like to circle the outer group of (5 in this case) points first. Then I put small lines through the median points in the three groups. I have circled the three points because we were asked to. My line went exactly through all three so I didn't have to move the line by the 'usual $\frac{1}{3}$ of the distance'.

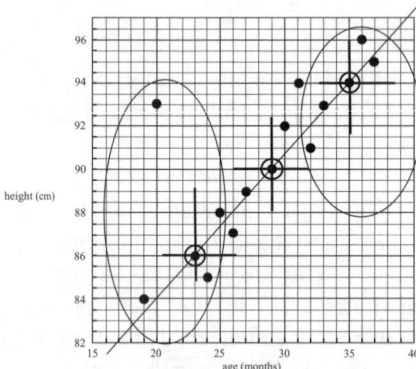

b. I will use the equation $y = a + bx$ to start with. The gradient is found by choosing two dots on the line. I chose (23, 86) and (35, 94).

$$\text{gradient} = \frac{y_2 - y_1}{x_2 - x_1}$$
$$= \frac{94 - 86}{35 - 23}$$
$$= \frac{8}{12} = 0.6666$$

So now we have $y = a + 0.6666x$.
Substitute (23, 86):
$$86 = a + 0.6666 \times 23$$
$$\therefore \quad a = 70.6682$$

The result: $y = a + 0.6666x$.
Now it is really important to convert x and y back to the correct variables names and to obey the 1 decimal place request.

The final answer is
$height = 70.7 + 0.7 \times age$. Note that there are other ways of setting this answer out.

c. In general, the use of the median is best when there are definite outliers. The point (20, 93) is very definitely an outlier.

Question 278

a. The histogram is asymmetrical with a positive skew.

b. i. There are six columns that indicate rainfalls of 500 mm or more. Adding the number of years from each column,
$$13 + 7 + 3 + 1 + 1 + 1 = 26 \text{ years.}$$

ii. Out of 103 years, this represents
$$\frac{26}{103} \times 100 = 19.4\%$$
(Correct to 1 decimal place.)

Question 279

a. Here, we are looking for the 'dot' which is lowest, or closest to the horizontal year-axis. It is the 1964 dot. Answer: 1964.

b. i. For the year 2010,
mean surface temperature
$$= -12.361 + 0.013 \times 2010$$
$$= 13.769$$
$$\approx 13.77°C$$

ii. Using the equation for 2000 again,
mean surface temperature
$$= -12.361 + 0.013 \times 2000$$
$$= 13.639°C$$
But the recorded temperature was 13.55°C.
The residual = actual – recorded
$$= 13.55 - 13.639$$
$$= 0.089$$
$$\approx 0.09°C$$

Comment. It is not a bad idea to calculate your answers to a higher degree of accuracy than the accuracy they ask for in the question. Round off in the last step.

204

Solutions: C1

iii. The gradient of the trend line (as given to us in the exam) is 0.013. That means that for each new year, the predicted temperature will rise by 0.013°C

Question 280

a. I have marked the missing point with a cross (×).

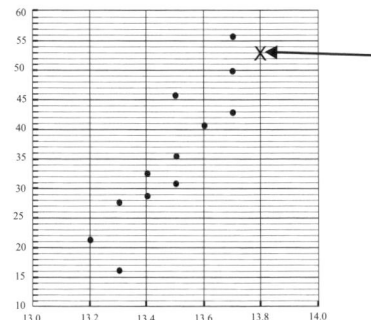

b. For this question, I typed the data from the table at the start of the question into my graphical calculator. For the moment, call the first column x, and the second column y.

i. The graphical calculator gave this result;
$y = a + bx$
$\therefore \quad y = -776.9323 + 60.25752x$
with a correlation coefficient of 0.9103. You will see why I calculated r when you get to part d. of the question.
Using the correct order of accuracy and using the correct variable names, the answer is

mean duration of warm spells
$= \quad -776.9 +$
$\quad \quad 60.3 \times$ mean surface temperature

ii. To plot this line, I calculated two coordinates, C and D. Using the equation $y = -776.9323 + 60.25752x$, when $x = 13.2$, $y = 18.47$ and when $x = 13.8$, $y = 54.62$.
That is, $C = (13.2, 18.47)$
$\quad \quad \quad D = (13.8, 54.62)$.
These are plotted on the following graph:

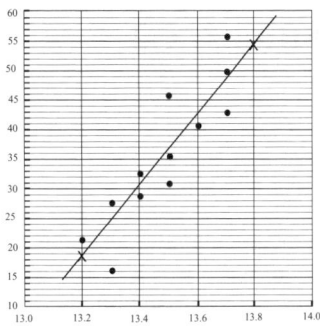

c. The residual plot does not support the idea of a curved graph. The number of points above and below the graph approximately equal. This is consistent with a linear fit being correct.

d. Earlier, I found that $r = 0.9103$
$\therefore \quad r^2 = 0.8286$
$\therefore \quad r^2 = 0.8286$
The percentage of variation in the mean duration of a warm spell that is explained by the variation in mean surface temperature is 82.85
$\approx \quad 83\%$

e. There is a strong positive, linear relationship between the mean duration of a warm spell and the mean surface temperature.

Solutions: C2

Question 281

a. One minute after pumping starts, the amount of water is
10 000 + 800 = 10 800 litres

b. Five minutes after pumping starts, the amount of water is
10 000 + 5(800) = 14 000 litres

c. $t_n = a + bn$
$\therefore\ t_n = 10\,000 + 800n$
So that $a = 10\,000$ and $b = 800$.

d. 20 000 = 800 × number of minutes
$\therefore$ The time taken is 25 minutes.

Question 282

a. The water decreases by 5% each week. This can be calculated by multiplying the previous week's value by 0.95. Thus at the end of the first week, there are 30 000 × 0.95 = 28 500 litres left.

b. At the end of the fourth week, there are $30\,000 \times 0.95^4 = 24\,435.1875$ or approximately 24 435 litres (correct to the nearest litre) left in the tank.

c. Method 1. Make a list in your calculator like this:

Week	Litres
0	30 000.00
1	28 500.00
2	27 075.00
3	25 721.25
4	24 435.19
5	23 213.43
6	22 052.76
7	20 950.12
8	19 902.61
9	18 907.48
10	17 962.11
11	17 064.00
12	16 210.80
13	15 400.26
14	14 630.25
15	13 898.74
16	13 203.80
17	12 543.61
18	11 916.43
19	11 320.61
20	10 754.58
21	10 216.85
22	9706.00

At the end of week 22, the water is below 10 000 litres.

Another method:
Level = $30\,000 \times 0.95^n$
$\therefore\ 10\,000 = 30\,000 \times 0.95^n$
$\therefore\ \dfrac{1}{3} = 0.95^n$

Either use 'trial and error' or maybe logs:
$\therefore\ \log\left(\dfrac{1}{3}\right) = n \log(0.95)$

$\therefore\ n = \dfrac{\log\left(\dfrac{1}{3}\right)}{\log(0.95)} = 21.4$ weeks

Which means you must wait until the end of week 22.

Question 283

a. $r = 1.10$ (the '10' because of the increase of 10%.). No, I suppose you don't really need the '0'…

b. After 5 weeks' usage = S_5

$= \dfrac{a(1-r^5)}{1-r}$

$= \dfrac{6000(1-1.10^5)}{1-1.10}$

$= 36\,630.6$

$\approx 36\,631$ litres (to the nearest litre).

Solutions: C2

c. Using the difference equation, I calculated that the Gerbers used 8000, 7490, 7881, 7424, ... litres in the first weeks. Thus in the third week, they used 7881 litres.

d. Here is a list of the week number and the water usage of both families:

Week	O'Callaghan	Gerber
1	6000.00	8000.00
2	6600.00	7940.00
3	7260.00	7881.20
4	7986.00	7823.58
5	8784.60	7767.10
6	9663.06	7711.76

Thus, in week 4, the O'Callaghans will use more water than the Gerbers.

Question 284

a. You could take either $\dfrac{2200}{2000}$ or $\dfrac{2420}{2200}$ to show that $r = 1.1$. (It would not be enough just to write 1.1 when it clearly asks you to *show* this.)

b. Multiplying by 1.1 is equivalent to increasing by 10%.

c. This is a geometric progression with $a = 2200$ and $r = 1.1$. Therefore,
$t_n = ar^{n-1}$
∴ $t_5 = ar^{5-1}$
$= 2000 \times 1.1^4$
$= 2928.4$
≈ 2928 heating units

Watch the units and watch the 'nearest whole number' requirement.

d. This is asking for the sum to ten terms of a GP.
$S_n = \dfrac{a(r^n - 1)}{r - 1}$
$= \dfrac{2000(1.1^{10} - 1)}{1.1 - 1}$
≈ 31874.8 heating units
(It doesn't specify an order of accuracy.)

e. We know that the second term is 2200 and the third term is 2420.
∴ $P_{n+1} = bP_n + c$ becomes
$P_{n+1} = bP_n + c$
∴ $P_2 = bP_1 + c$
∴ $2200 = b(2000) + c$ ... 1
and $2420 = b(2200) + c$ ... 2
2−1 $220 = 200b$
∴ $b = 1.1$ and from 1,
$2200 = (1.1)(2000) + c$
∴ $c = 0$
Answer: $b = 1.1$ and $c = 0$.

Question 285

a. Interpret the information as follows: the cost of n outlets is $C_n = 3500 + 80n$ where n is the number of outlets beyond 5. For 8 outlets, the cost will be $3500 + 80(3) = \$3740$.

b. Let $C_n = 3500 + 80n = 4400$
∴ $3500 + 80n = 4400$
∴ $80n = 4400 - 3500$
∴ $n = \dfrac{900}{80}$
∴ $n = 11.25$

That may be interpreted as: 11 additional units may be purchased with some change, but 12 units can't quite be afforded. Answer: Buy an additional 11 outlets which means 16 outlets in total.

c. If 20 squares means 12 outlets, then for 35 squares of living area,
$\dfrac{20}{12} = \dfrac{35}{x}$
∴ $20x = 12 \times 35$
∴ $x = \dfrac{420}{20}$
$= 21$ outlets

To find the cost, remember that the cost equation needs to know how many additional outlets (beyond 5) will be

207

purchased. In this case 21 − 5 = 16 additional units.
∴ $C_n = 3500 + 80n$
= $C_n = 3500 + 80 \times 16$
= $4780

Question 286

$S_n = 1.2S_{n-1} - 200$. The number, S_n, of heating systems sold in the nth year is generated by the difference equation where $n \leq 5$ and $S_3 = 2224$.

a. We are given S_3 so we will need to find S_2 before we can find S_1.
$S_n = 1.2S_{n-1} - 200$
∴ $S_3 = 1.2S_2 - 200$
∴ $2224 = 1.2S_2 - 200$
∴ $2424 = 1.2S_2$
∴ $S_2 = \dfrac{2424}{1.2} = 2020$

and
$S_2 = 1.2S_1 - 200$
∴ $2020 = 1.2S_1 - 200$
∴ $2220 = 1.2S_1$
∴ $S_2 = \dfrac{2220}{1.2} = 1850$ heating systems.

b. The number of heating units **sold** in the first 3 years:
$1850 + 2020 + 2224 = 6094$.
For the number made, we need to refer back to the table in Question 309.

The number will be
$2000 + 2200 + 2420 = 6620$
The required percentage is
$\dfrac{6094}{6620} \times \dfrac{100}{1}$
= 92.054
≈ 92.1%
Watch that order of accuracy requirement.

Question 287

a. This is an Arithmetic Progression with $a = 4.2$ and $d = 0.3$. Term number 5 is

$t_5 = a + 4d$
= $4.2 + 4(0.3)$
= 5.4 km.

b. We want
$t_n = a + (n-1)d = 7$
∴ $t_n = 4.2 + (n-1)(0.3) = 7$
∴ $4.2 + 0.3n - 0.3 = 7$
∴ $0.3n = 7 - 4.2 + 0.3$
∴ $0.3n = 3.1$
∴ $n = 10.33$
Now, you can't really have $n = 10.33$ and $n = 10$ would not be enough terms. Thus, Day 11.

c. At the start of Day 3, only two distances have been run. The sum of the first two terms:
$S_n = \dfrac{n}{2}(2a + (n-1)d)$
$S_2 = \dfrac{2}{2}(2(4.2) + (2-1)(0.3))$
= $(8.4 + 0.3) = 8.7$ km.
At the end of Day 12, 12 distances have been run.
$S_{12} = \dfrac{12}{2}(2(4.2) + (12-1)(0.3))$
= $6(8.4 + 11 \times 0.3)$
= 70.2 km
The difference between these two is the inclusive distance run:
$70.2 - 8.7 = 61.5$ km

d. $t_n = a + (n-1)d$
∴ $t_n = 4.2 + (n-1)(0.3)$
= $4.2 + 0.3n - 0.3$
= $3.9 + 0.3n$
= $0.3n + 3.9$
This looks like $0.3n + b$ if $b = 3.9$.

Question 288

a. Consider the terms 3, 3.5, 4.05. The two differences are 0.5 and 0.55. They are not equal so the terms are not in an Arithmetic Progression. The two

Solutions: C2

consecutive ratios are $\frac{3.5}{3}$ and $\frac{4.05}{3.5}$. Since these are not equal, the terms are not in a Geometric Progression.

b. To find the second term,
$$w_2 = 1.1w_1 + c$$
$$\therefore 3.5 = 1.1(3) + c$$
$$\therefore c = 3.5 - 3.3$$
$$= 0.2$$

c. Use the equation $w_{n+1} = 1.1w_n + 0.2$.
The distance walked on Day 4:
$$w_{n+1} = 1.1w_n + 0.2$$
$$\therefore w_4 = 1.1w_3 + 0.2$$
$$= 1.1(4.05) + 0.2$$
$$= 4.655 \text{ km.}$$
The distance walked on Day 5:
$$w_5 = 1.1w_4 + 0.2$$
$$= 1.1(4.655) + 0.2$$
$$= 5.322 \text{ km.}$$
The question asks for two decimal places. Therefore it is wise to work to three or more decimal places. Answer: 5.32 km

Question 289

The table shows a geometric progression with $a = 10$ and $r = 1.05$.

a. The second term is $10 \times 1.05 = 10.5$.

b. The total distance is the sum of 14 terms:
$$S_n = \frac{a(r^n - 1)}{r - 1}$$
$$\therefore S_{14} = \frac{10(1.05^{14} - 1)}{1.05 - 1}$$
$$= \frac{10(1.05^{14} - 1)}{0.05}$$
$$= 195.9863$$
$$\approx 196 \text{ km (to the nearest km.)}$$

c. i. The two parts are $\frac{2}{5}$ and $\frac{3}{5}$. The walking part is $\frac{2}{5} \times 10 = 4$ km.

ii. She runs $\frac{1}{4}$ of the distance and walks $\frac{3}{4}$ of the distance. That makes the ratio 1:3.

d. i. For the difference equation, the next term is found by multiplying the last term by 0.9 and then adding 1.2:
$$t_{n+1} = 0.9t_n + 1.2 \quad \text{where } t_1 = 5.$$

ii. One way of solving this is to convince your calculator to list the first 30 or so terms of the progression. You could use the difference equation section of your calculator, for example. After about 30 terms, you start to see that they have values approaching 12 – but they don't actually reach 12.
You could test that using algebra. If the current term is 12, then the next term is
$$t_{n+1} = 0.9t_n + 1.2$$
$$\therefore t_{n+1} = 0.9(12) + 1.2$$
$$= 10.8 + 1.2 = 12$$
Which is quite similar. The terms don't go beyond 12.

Question 290

a. Start with 48 000 kg and pick two lots of 3000, and you will be left with 42 000 kg of fruit.

b. $F_n = 48000 + d \times n$. We lose 3000 kg of fruit per day. Thus d must be -3000. You can test this by trying $n = 2$ and comparing the result to the first answer. Watch that '–' sign.

c. We want $F_n = 0$
$$\therefore 0 = 48000 - 3000 \times n$$
$$\therefore n = \frac{48000}{3000} = 16 \text{ days.}$$

Question 291

a. The terms of the progression are: 625, 500, 400, ...

Solutions: C2

The value of r is found by taking any term and dividing by the term before it.
$\frac{500}{625} = 0.8$ as required.

b. $t_n = a \times r^{n-1}$
∴ $t_5 = 625 \times 0.8^4$
  $= 256$ hours.
Alternatively you could simply take 400 and multiply by 0.8 and then multiply the result by 0.8.

c. The result of $t_n = 625 \times 0.8^{n-1}$ should be re-written as $H_5 = 625 \times 0.8^{n-1}$. Note that it does not seem reasonable to write the answer as a difference equation such as $H_n = H_{n-1} \times 0.8$.

d. We will need to find the hours worked in the 6th month and subtract the hours worked in the 7th month.
$t_6 - t_7$
$= ar^{6-1} - ar^{7-1}$
$= 625(0.8^{6-1} - 0.8r^{7-1})$
$= 40.96 \approx 41$ hours.

e. Rather than treat this as an inequality, ask in what month the number of hours will be equal to 100.
$ar^{n-1} = 100$
∴ $625 \times 0.8^{n-1} = 100$
∴ $0.8^{n-1} = 0.16$
Either use logs or a graphical calculator method.
∴ $\log(0.8^{n-1}) = \log(0.16)$
∴ $(n-1)\log(0.8) = \log(0.16)$
∴ $n - 1 = \frac{\log(0.16)}{\log(0.8)}$
∴ $n = 1 + \frac{\log(0.16)}{\log(0.8)}$
$= 9.2126$
We are after a whole number of months. Working longer than 9.2126 months will ensure that less than 100 hours will be needed.
Answer: During the 10th month.

f. We will need to find the sum of the hours worked in the 12 months and subtract 1525. Note that the 'next nine months' takes us to the 12th month.
$S_{12} - 1525$
$= \frac{a(1-r^n)}{1-r} - 1525$
$= \frac{625(1-0.8^{12})}{1-0.8} - 1525$
$= \frac{625(1-0.8^{12})}{0.2} - 1525$
$= 2910.25 - 1525$
$= 1385.25 \approx 1385$ hours.

Question 292

a. The equation $V_{n+1} = rV_n + d$ must become 'take the previous volume, reduce it by 10% (a multiplication factor of 0.9) and add 2000'. Thus, $r = 0.9$ and $d = 2000$.

b. Here is a list of the volumes on the first 4 days:

n	V
1	45 000
2	42 500
3	40 250
4	38 225

Answer: 38 225 litres.

c. Here is a list of the volumes over enough days to see 30 000 or less to appear. It shows day 10.

n	V
1	45 000
2	42 500
3	40 250
4	38 225

Solutions: C2

5	36 402.5
6	34 762.25
7	33 286.03
8	30 761.6
9	29 685.51

d. The terms 2000, 2000 × 0.9, 2000 × 0.9²... form a geometric progression of the added water. They add to

$$S_\infty = \frac{a}{1-r}$$
$$= \frac{2000}{1-0.9}$$
$$= 20\,000 \text{ litres.}$$

If you want to see a different way of achieving this, look at the 119th term. It is 20 000.1 litres. Not during the exam though...

Question 293

a. From the graph, Maria's kilojoule intake on day 1 was 8700 kj.

b. If you were to join the dots with a straight line, its gradient would be
$$\frac{8100 - 8700}{4} = -150.$$
That is, there is a drop of 150 kj each day. On day six, we would expect the day 5 value to drop by 150.
$8100 - 150 = 7950 \text{ kj.}$

c. We know that on day 1 ($n = 1$), the kilojoule intake was 8700.
∴ $8700 = a - 150 \times (1)$
∴ $8700 + 150 = a$
∴ $a = 8850$

d. Now we have the equation $K_n = 8850 - 150 \times n$ so that when the intake is 6750,
$6750 = 8850 - 150 \times n$
∴ $150 \times n = 8850 - 6750$
∴ $150 \times n = 2100$

∴ $n = \frac{2100}{150}$
∴ $150 \times n = 2100$
∴ $n = 14$

Maria's daily kilojoule intake be 6750 kj on the 14th day.

Question 294

a. The common ratio is 0.95. That is, each term drops in value by 0.05 to form the next term. Rupert's kilojoule intake is reduced by 5% each day

b. The third term is
$12\,000 \times 0.95 \times 0.95$
$= 12\,000 \times 0.95^2$
$= 10\,830 \text{ kj.}$

c. Using the pattern from part b., we can predict that on the nth day Rupert's kilojoule intake is
$R_n = 12\,000 \times 0.95^{n-1}$

d. On day 10, Rupert's kilojoule intake is
$R_{10} = 12\,000 \times 0.95^9$
$= 7562.99$
On day 9, Rupert's kilojoule intake was
$R_9 = 12\,000 \times 0.95^8$
$= 7961.05.$
The difference is $7961.05 - 7562.99$
$= 398.06$
$\approx 396 \text{ kj.}$

e. Use the formula for the sum of terms in a G.P.
$$S_n = \frac{a(1-r^n)}{1-r}$$
By day 14, Rupert's total kilojoule intake was
$$S_{14} = \frac{12\,000(1-0.95^{14})}{1-0.95}$$
$= 122\,958.01 \text{ kj.}$
By day 7, Rupert's total kilojoule intake was

Solutions: C2

$$S_7 = \frac{12\,000(1-0.95^7)}{1-0.95}$$
$= 72\,399.05$ kj.
Rupert's total kilojoule intake in the period from day 8 to day 14 is
$122\,958.01 - 72\,399.05$
$= 50\,558.96$
$\approx 50\,559$ kj.

Question 295

a. The question is asking for the fourth term. We use the difference equation
$M_{n+1} = 0.75 M_n + 8$.
$M_2 = 20$
$\therefore M_3 = 0.75 M_2 + 8$
$= 0.75 \times 20 + 8$
$= 23$.
Then,
$M_4 = 0.75 M_3 + 8$
$= 0.75 \times 23 + 8$
$= 25.25$ minutes.

b. We are asked about the progression
20, 23, 25.5, ...
To test whether it is an arithmetic progression, choose any term (beyond the first term) and subtract the term before it. Repeat that process to see if the answer is the same.
$23 - 20 = 3$
$25.5 - 23 = 2.5$
Since these two answers are not the same, it is not an arithmetic progression.
To test whether it is a geometric progression, choose any term (beyond the first term) and divide by the term before it. Repeat that process to see if the answer is the same.
$\frac{23}{20} = 1.15$
$\frac{25.5}{23} = 1.11$
Since these two answers are not the same, it is not a geometric progression either.

c. To find out how many minutes Maria will cycle on day 1,

$M_2 = 0.75 M_1 + 8$
$\therefore 20 = 0.75 M_1 + 8$
$\therefore 12 = 0.75 M_1$
$\therefore M_1 = \frac{12}{0.75}$
$= 16$ minutes.

Question 296

Consider the progression of terms from Rupert's swimming:
100, 150, 200, ...
this is an arithmetic progression with $a = 100$ and $d = 50$. The equation for t_n is
$t_n = 100 + (n-1)50$.
Now consider the progression of terms from Rupert's running:
$500, 500 \times 1.02, 500 \times 1.02^2 ...$
This is a geometric progression with $a = 500$ and $r = 1.02$. The equation for t_n is
$t_n = 500 \times 1.02^{n-1}$.
We wish to find when
$100 + (n-1)50 > 500 \times 1.02^{n-1}$.
In my graphical calculator, I created two equations, $y_1 = 100 + (x-1)*50$ and $y_2 = 100 \times 1.02^{(x-1)}$ and drew their graphs. I also looked at the table of values each generated. Here is a section of it.

n	Swimming distance	Running distance
10	550	697
11	600	609
12	650	621
13	700	634

On day 11, the swimming distance is less than the running distance. But on day 12, for the first time the swimming distance is greater than the running distance.
Answer: Day 12.

Solutions: C3

Question 297

a. $\cos\theta = \dfrac{8}{20}$

$\therefore\ \theta = 66.42182° \approx 66°$

b. The diagram now looks like this:

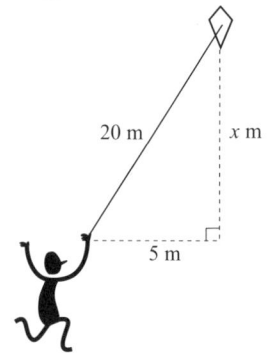

By Pythagoras:
$20^2 = 5^2 + x^2$
$\therefore\ x^2 = 400 - 25$
$\therefore\ x^2 = \sqrt{375}$
$= 19.36492$
≈ 19 m

Question 298

a. The triangle looks like this:

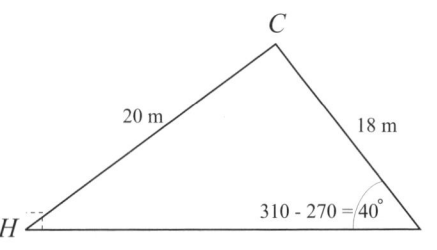

$\angle HDC = 40°$

b. We can use the sine rule to find $\angle CHD$:

$\dfrac{20}{\sin 40°} = \dfrac{18}{\sin \angle CHD}$

$\therefore\ \sin \angle CHD = \dfrac{18 \sin 40°}{20}$

$= 0.578509$

$\therefore\ \angle CHD = 35.34573$

$\approx 35.35°$

Question 299

a. If you turn the kite on its side and redraw triangle *ABC*, it looks like this:

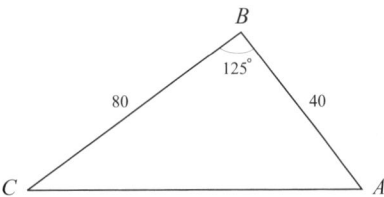

To find the length AC, we need the cosine rule:
$AC^2 = 80^2 + 40^2 - 2(80)(40)\cos 125°$
$= 8000 + 3670.89$ (do it in 2 stages...)
$= 11670.89$
$\therefore\ AC = 108.003$
≈ 108 cm

b. To find the area of the kite, we should find the area of the triangle *ABC* and double it. Since we have the lengths of two sides and the included angle, we use the formula

$A = \dfrac{1}{2} bc \sin A$

$= \dfrac{1}{2}(80)(40)\sin(125°)$

$= 1310.643$

Thus, the whole kite has area 2×1310.643
$= 2621.287$
≈ 2621 cm^2 as required

c. To find $\angle ABE$, first find $\angle CAB$. Using the sine rule:

$\dfrac{80}{\sin \angle CAB} = \dfrac{AC}{\sin 125°}$

213

Solutions: C3

∴ $\sin \angle CAB = \dfrac{80 \sin 125°}{108.03}$

= 37.34°

≈ 37°

From $\triangle ABC$, $\angle ABE = 180 - 90 - 37$

= 53°

d. The lengths ratio is 2:3 so that the area ratio is 4:9. Thus, the required area is $2621 \times \dfrac{9}{4} \approx 5897 \text{ cm}^2$.

Question 300

a. Here is the triangle in question. Use the area formula $A = \dfrac{1}{2}bc \sin A$:

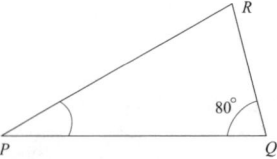

$= \dfrac{1}{2}(1203)(951)\sin(80°)$

≈ 563 336 m²

b. To find the distance from P to R, use the cosine rule:

$PR^2 = 1203^2 + 951^2 - 2(1203)(951)\cos 80°$

∴ $PR \approx 1398$ m.

c. From P to R? You really need to consider this triangle:

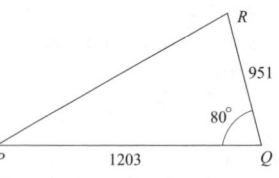

We need to find the angle QPR. It is found by using the sine rule, and using the answer from part **b**.

$\dfrac{951}{\sin \angle QPR} = \dfrac{PR}{\sin 80°}$

∴ $\sin \angle QPR = \dfrac{951 \times \sin 80°}{1398}$

= 0.6699

∴ $\angle QPR = 42.06°$

Now, the bearing from P to R is:

90 − 42.06

= 47.94

≈ 049° T

d. For this question, we will need to find the angle SPR first. To do this start by piecing together the 15° bearing that is hiding in the sentence at the start of the question along with the 42.06° we found in the last part:

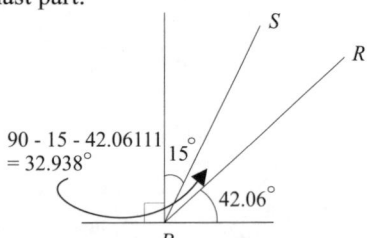

90 - 15 - 42.06111
= 32.938°

Using this result:

$A = \dfrac{1}{2}bc \sin A$

$= \dfrac{1}{2}(1048)(1398)\sin(32.938°)$

≈ 398 310 m²

(398 440 if you only used 2 decimal places)

e. To find the area of the camping ground, add the two areas we found (part **a.** plus part **d.**):

563 336 + 398 310

= 961 650 m²

Question 301

a. Using Pythagoras' theorem,

$BC^2 = 2.20^2 + 8.21^2$

= 72.2441

∴ $BC = 8.49965$

≈ 8.50 m   (two decimal places)

b. $\tan(\angle ACB) = \dfrac{2.20}{8.21}$

= 0.267965

∴ $\angle ACB = 15°$

214

Solutions: C3

Question 302

a. The total angle in the centre of this shape is 360°.

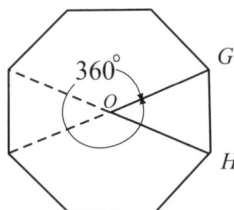

Therefore, $\angle GOH = \dfrac{360}{8} = 45°$.

b. The formula needed here is:
$$\text{Area}_{\text{triangle}} = \dfrac{1}{2} bc \sin(A)$$
$$= \dfrac{1}{2}(2.30)(2.30)\sin(45°)$$

On my calculator, this is 1.870297. Now there are 8 of these sections, so the total area is 14.96238, or to the nearest square metre, 15 m². Hmmm, maybe I used too many decimal places...

c. For the side length GH, we need to consider the triangle OGH:

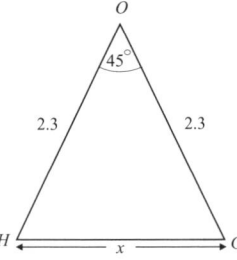

Using the cosine rule,
$x^2 = 2.3^2 + 2.3^2 - 2(2.3)(2.3)\cos 45°$
$= 10.58 - 7.4812$ (do it in 2 stages...)
$= 3.0988$
$\therefore \quad x = 1.7603$
≈ 1.76 m

Note: dividing the triangle down the centre is another method. It allows conventional trigonometry to be used.

d. i. Here is the fastest method I found. Create a triangle from G to the centre to K and back again:

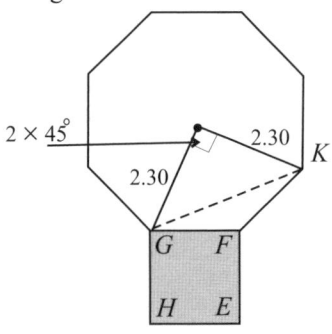

$GK^2 = 2.30^2 + 2.30^2$
$= 10.58$
$\therefore \quad GK = 3.2569$
≈ 3.25 m

ii. Length OK is simply length GK with 2.30 m added to it. Thus, 5.55 m.

e. The length of PR is found using the sine rule:
$$\dfrac{PR}{\sin(105°)} = \dfrac{3}{\sin(35°)}$$
$$\therefore \quad PR = \dfrac{3 \times \sin(105°)}{\sin(35°)}$$
$= 5.0521$ m
≈ 5.05 m

f. It is necessary to look closely at the shape $YQRZ$ and to locate the 2.7 metres in two places:

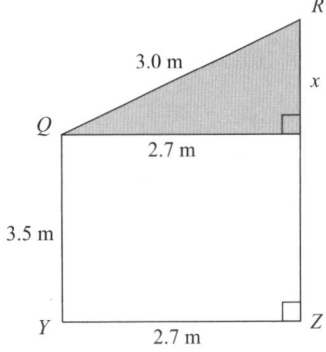

Solutions: C3

From the shaded triangle, we can use Pythagoras' theorem
$$3^2 = 2.7^2 + x^2$$
$$\therefore \quad x^2 = 1.71$$
$$\therefore \quad 1.3077 \text{ m}$$
Now, the length of pole RZ is that answer plus the 3.5 m of the other poles.
$$1.3077 + 3.5$$
$$= 4.8077 \text{ m}$$
$$= 480.77 \text{ cm}$$
$$\approx 481 \text{ cm}$$

Question 303

a. The volume of a cone is given by the formula
$$V_{cone} = \frac{1}{3}\pi r^2 h$$
$$= \frac{1}{3}\pi(0.6)^2(0.7)$$
$$= 0.263893 \approx 0.26 \text{ m}^3$$

b. The dimensions of this cone are twice the dimensions of the first.
$$\frac{V_{large}}{V_{small}} = \left(\frac{\text{length}_{large}}{\text{length}_{small}}\right)^3$$
$$\frac{V_{large}}{V_{small}} = \left(\frac{2.4}{1.2}\right)^3$$
$$\therefore \quad \frac{V_{large}}{V_{small}} = (2)^3 = 8$$
There is a 1:8 ratio in volumes.

Question 304

a. Angle ABC is marked here as θ:

$$\tan\theta = \frac{10}{3.6}$$
$$\therefore \quad \theta = \tan^{-1}\frac{10}{3.6}$$
$$= 70.2°$$
$$\approx 70°$$
Note: the question says to write your answer correct to the nearest degree.

b. From the same diagram, Pythagoras gives:
$$BC^2 = 3.6^2 + 10^2$$
$$= 112.96$$
$$\therefore \quad BC = \sqrt{112.96}$$
$$= 10.63$$
$$\approx 10.6 \text{ m}$$
Remember units and watch that 'order of accuracy' requirement.

c. We will be using this diagram:

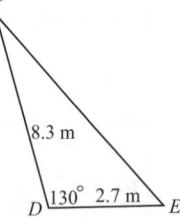

The cosine rule:
$$EF^2 = 8.3^2 + 2.7^2 - 2(8.3)(2.7)\cos(130°)$$
$$= 76.18 - (-28.8097)$$
$$= 104.9897$$
$$\therefore \quad EF = 10.24645$$
$$\approx 10.2 \text{ metres (one decimal place)}$$

d. For the area of the sail DEF.
$$A = \frac{1}{2}(8.3)(2.7)\sin(130°)$$
$$= 8.583528$$
$$\approx 8.6 \text{ square metres}$$
Note: Heron's formula could be used, but if you do, it is best to use the higher accuracy answer from part **c**.

Question 305

a. Begin by extending the diagram:

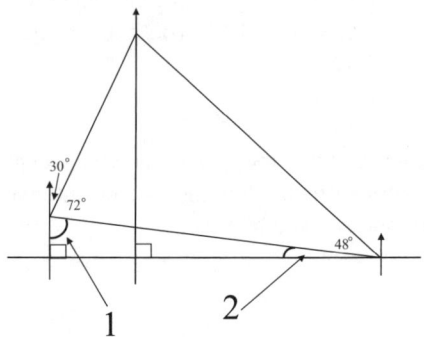

Solutions: C3

Start by finding the angle marked 1. It is on a straight line with 30° and 72°. Therefore it is (180 – 30 – 72)° = 78°.
Now, angle 2 is (90 – 78)° = 12°.
Since this is from the west line, the bearing of V from U is $(270 + 12) = 282°T$.

b. The distance TU may be found using the sine rule:
$$\frac{TU}{\sin(72°)} = \frac{5.4}{\sin(48°)}$$
$$\therefore TU = \frac{5.4 \times \sin(72°)}{\sin(48°)}$$
$$= 6.09177$$
$$\approx 6.9 \text{ km (correct to 1 decimal place)}$$

c. To find the shortest distance to complete the race, add all three side lengths. We know two of them already, but it is necessary to find length VU now. It can be found either by the sine rule or the cosine rule.
Now, angle VTU is $180 - (72 + 48) = 60°$ and so the sine rule becomes:
$$\frac{VU}{\sin(60°)} = \frac{5.4}{\sin(48°)}$$
$$\therefore VU = \frac{5.4 \times \sin(60°)}{\sin(48°)}$$
$$= 6.2929$$
$$\approx 6.3 \text{ km (correct to 1 decimal place)}$$
Now add the other two lengths, 5.4 and 6.9 (you could use the 'higher accuracy' 6.0977 but 6.9 is OK).
$$= 5.4 + 6.9 + 6.3 = 18.6 \text{ km}$$

Question 306

a. Since triangle XYZ is an equilateral triangle, all interior angles are 60°.

b. The distance required here was conveniently marked with a dotted line.
Note that an angle of 30° is created for $\angle OXA$:

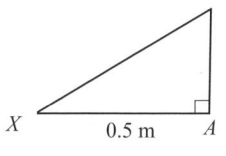

$$\tan 30° = \frac{\text{opposite}}{\text{adjacent}}$$
$$\therefore 0.5773 = \frac{OA}{0.5}$$
$$\therefore OA = 0.5 \times 0.5773$$
$$= 0.2886 \approx 0.289 \text{ metres}$$

c. The area of the circle is πr^2. The area of the whole triangle is
$$\frac{1}{2} \times 1 \times 1 \times \sin 60° = 0.4330. \text{ Thus,}$$
$$\therefore \frac{0.4330 - \pi r^2}{\pi r^2} = \frac{2}{1}$$
$$\therefore 2\pi r^2 = 0.4330 - \pi r^2$$
$$\therefore 3\pi r^2 = 0.4330$$
$$\therefore r^2 = \frac{0.4330}{3\pi}$$
$$= 0.045943$$
$$\therefore r = 0.214343$$
$$\approx 0.214 \text{ metres}$$

Question 307

a. H has height 200. W has height 150. The difference is 50 m.

b. This is a 3-dimensional problem. The difference in height is 100 - 50 = 50 m. The horizontal distance is 500 metres as shown here. Therefore the distance between them is:
$$= \sqrt{100^2 + 500^2}$$
$$= \sqrt{260000}$$
$$= 509.902$$
$$\approx 510 \text{ m correct to the nearest metre.}$$

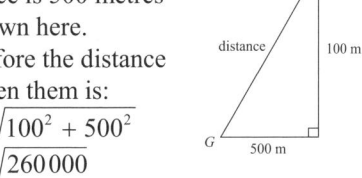

c. Given: $\frac{\text{height}}{\text{width}} = 0.12$. The difference in height between them is 150 m.
$$\therefore \frac{150}{\text{width}} = 0.12$$
$$\therefore \frac{150}{0.12} = \text{width}$$
$$\therefore \text{width} = 1250 \text{ m}.$$

217

Solutions: C3

d. $\dfrac{\text{real world}}{\text{map measurement}} = 40\,000$

∴ $\dfrac{2 \times 1000 \times 100}{\text{map measurement}} = 40\,000$

∴ map measurement $= \dfrac{200\,000}{40\,000}$

$= 5$ cm.

Question 308

a. i. The diagram can be simplified for this question:

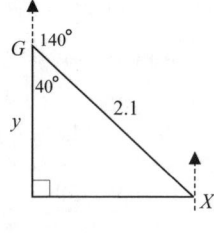

$\cos(40°) = \dfrac{y}{2.1}$

$y = 2.1\cos(40°)$

$= 1.609$

≈ 1.6 km

ii. Start by finding the distance North of Y from X. The diagram looks like this:

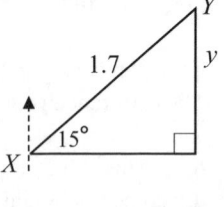

$\sin(15°) = \dfrac{y}{1.7}$

$y = 1.7\sin(15°)$

$= 0.4400$

So we take the previous answer and subtract 0.4400:

$1.609 - 0.4400$

$= 1.16$ km ≈ 1.2 km.

b. To find the size of angle GXY, construct the lines shown here:

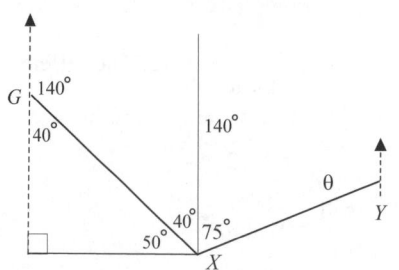

Note the 40° under the 140°, and the use of 'angles in a triangle add to 180°'. The required angle, $\angle GXY$, is $40°+75° = 115°$.

c. Use the cosine rule with triangle GXY:

$GY^2 = 2.1^2 + 1.7^2 - 2(2.1)(1.7)\cos(115°)$

$= 7.3 + 3.01749$

$= 10.31749$

∴ $GY = 3.212086$

≈ 3.2 km.

d. The required angle (θ) is shown here. To find that angle, first find angle XGY using the cosine rule:

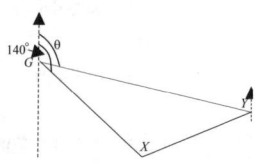

$1.7^2 = 2.1^2 + 3.2^2 - 2(2.1)(3.2)\cos(XGY°)$

∴ $-11.76 = -13.44\cos(XGY°)$

∴ $\cos(XGY°) = \dfrac{11.76}{13.44}$

∴ $XGY = 28.955°$

Now we have a 'sine rule' problem.

$\dfrac{\sin(XGY)}{1.7} = \dfrac{\sin(115°)}{3.2}$

∴ $\sin(XGY) = \dfrac{\sin(115°)}{3.2} \times 1.7$

$= 28.7818°$

Thus, the required bearing is $140 - 28.7818$

$= 111.2182°$

$\approx 111°$

Question 309

a. Construct a suitable triangle inside the main diagram:

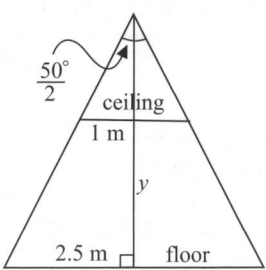

Solutions: C3

From the triangle associated with the ceiling,
$$\tan(25°) = \frac{1}{\text{height}_{small}}$$
$$\therefore \text{height}_{small} = \frac{1}{\tan(25°)}$$
$$= 2.144507 \text{ m}.$$

From the triangle associated with the floor,
$$\tan(25°) = \frac{2.5}{\text{height}_{small}}$$
$$\therefore \text{height}_{big} = 5.361267 \text{ m}.$$

The difference between them is the required height:
$$5.361267 - 2.144507$$
$$= 3.21676$$
$$\approx 3.2 \text{ m}$$
Correct to 1 decimal place.

b. It is only necessary to consider the end of the prism, since the lengths from the front to the back are the same.
The ratio of the areas will be the ratio of two corresponding lengths squared. One length is the 5 metres on the larger triangle and the other is the 5 metres on the smaller triangle
$$\frac{MNC \text{ area}}{ABC \text{ area}} = \left(\frac{2}{5}\right)^2$$
$$\therefore \frac{4}{25}$$
That is the answer – the fraction of the upper triangle area compared to the lower triangle area.

Question 310

a. Triangle AXY is a right-angled triangle.
$$\therefore \angle AXY = 90 - 45°$$
$$= 45°$$

b. From the right-angled triangle AXY:
$$\cos 45° = \frac{55}{AX}$$

$$\therefore AX = \frac{55}{\cos 45°}$$
$$= 77.7817$$
$$\approx 77.8 \text{ metres}.$$
Alternatively, you could use Pythagoras' theorem, knowing that length AY and length XY are equal in magnitude.

c. A further diagram is needed, with the bearings made more apparent:

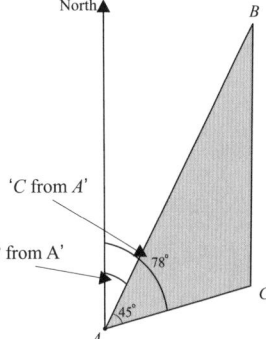

The required angle 'B from A' is
$78° - 45° = 33°$.
As a bearing, this is $033°$.

d. One approach is to use the Cosine rule with triangle AXC:
$$a^2 = b^2 + c^2 - 2bc\cos A$$
$$\therefore XC^2 = AX^2 + AC^2 - 2XC.AC\cos A$$
$$= 77.8^2 + 142^2 - 2(77.8)(142)\cos 45°$$
$$= 26216.84 - 15623.67$$
$$= 10593.17$$
$$\therefore XC = \sqrt{10593.17}$$
$$= 102.923 \approx 102.9 \text{ metres}.$$
(Another method: use right-angled triangle XYC.)

e. Use the area formula:
$$A = \frac{1}{2}bc\cos A$$
$$= \frac{1}{2}AC \times AB \times \sin(45°)$$
$$= \frac{1}{2}(142)(251)\sin(45°)$$
$$= 12601.34995 \approx 12\,601 \text{ square metres}.$$

f. **i.** Use the Cosine rule with triangle ABC:
$$a^2 = b^2 + c^2 - 2bc\cos A$$
$$\therefore$$
$$BC^2 = AB^2 + AC^2 - 2AB.AC\cos(45°)$$

Solutions: C3

$$= 251^2 + 142^2 - 2(251)(142)\cos 45°$$
$$= 83\,165 - 50\,405.40$$
$$= 32\,759.60$$
$$\therefore XC = \sqrt{32759.60}$$
$$= 181.0 \text{ metres}$$

Oops, I probably should have stopped ages ago. They only asked for 'how this ... may be found'.

ii. Use the sine rule:
$$\frac{a}{\sin A} = \frac{b}{\sin B}$$
$$\therefore \frac{142}{\sin \angle ABC} = \frac{181}{\sin 45°}$$
$$\therefore 181\sin \angle ABC = 142\sin 45°$$
$$\therefore \sin \angle ABC = \frac{142\sin 45°}{181} = 0.55475$$
$$\therefore \angle ABC = 33.693°$$
$$\approx \angle ABC = 33.7°$$

g. Note that because of the 45°, lengths AN and MN are equal (x).
Use the given area as follows:
$$A = \frac{1}{2} \times \text{base} \times \text{height}$$
$$\therefore 3200 = \frac{1}{2} \times x \times x$$
$$\therefore 6400 = x^2$$
$$\therefore x = 80 \text{ metres}.$$

Question 311

a.

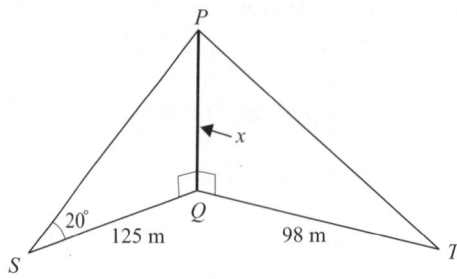

Call the height of the tower x. From the diagram,
$$\tan 20° = \frac{x}{125}$$

$$\therefore x = 125\tan 20°$$
$$= 45.4963$$
$$\approx 45.5 \text{ metres}.$$

b. Start by finding angle QPT. From triangle QPT,
$$\tan \angle QPT = \frac{98}{x}$$
$$= \frac{98}{45.5}$$
Use inverse tan:
$$\angle QPT = 65.10°$$
The angle of depression is marked here:

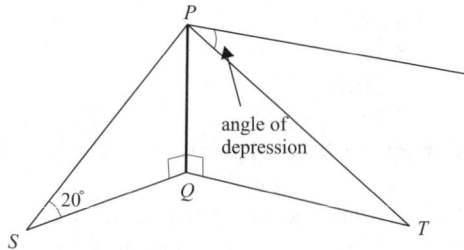

The required angle is $90° - \angle QPT$
$$= 90° - 65.10°$$
$$\approx 24.9°$$
Note: another method might have been to find angle QTP.

Question 312

a. The outside radius of the tank is $\frac{3.5}{2} = 1.75$ metres. The inside radius is found by subtracting 0.25 metres from this. The result is 1.5 metres.

b. Water fills the inside tank. This is a cylinder of radius 1.5 metres and height $2.4 - 0.25 - 0.25 = 1.9$ metres.
$$V_{cylinder} = \pi r^2 h$$
$$= \pi \times 1.5^2 \times 1.9$$
$$= 13.43$$
$$\approx 13 \text{ cubic metres}.$$

Solutions: C3

Question 313

a. W is halfway between Q and R. Thus,
$$QW = \frac{1}{2} \times 24$$
$$= 12 \text{ cm}.$$

b. From the right angled triangle, WAQ, we can find that
$$\tan(\angle WAQ) = \frac{12}{32}.$$
On your calculator, press $\tan^{-1}\left(\frac{12}{32}\right)$
$$= 20.556$$
$$\approx 20.6°.$$

c. The required angle $\angle AWB$ can be found by adding angles A and B. I have put that idea on the diagram below.

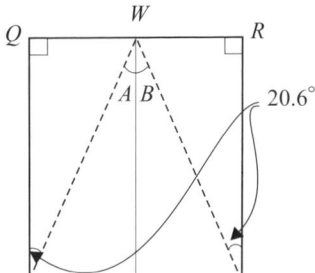

Each of A and B is $20.6°$ (they are 'alternate angles'). Thus,
$$\angle AWB = 2 \times 20.556$$
$$= 41.112$$
$$\approx 41.1°$$

Comment. It is a good idea to go back to a higher degree of accuracy whenever possible. That is, $20.556°$ instead of $20.6°$.

d. For this part of the question, I note that only one mark has been allocated. Triangle AWB has exactly half the area of rectangle $ABRQ$. You could simply write that result down to earn your mark.
For those of you who wish to know a bit more about why it is one half, consider the four triangles in the following diagram. They are all identical.

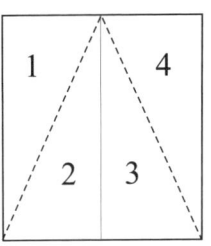

Question 314

a. Assuming that the front face of Tessa's carving is an isosceles triangle, you can form a right angle triangle. it is shaded in the following diagram.

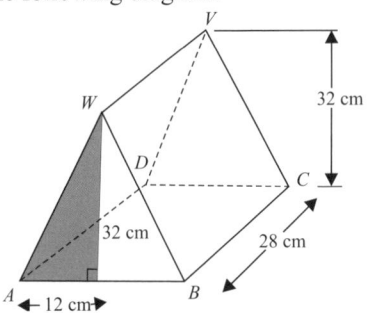

The two known lengths of this triangle are 12cm and 32 cm.

Using Pythagoras' theorem,
$$AW^2 = 12^2 + 34^2$$
$$\therefore AW = \sqrt{32^2 + 12^2}$$
$$= 34.176$$
$$\approx 34 \text{ cm}.$$

b. We are being asked for the total outside surface area of this shape. Identify five different areas. Two of them are the sloping rectangles on the sides. One area is from the rectangle underneath, and the remaining two are the triangles on the front and back.
Underneath, the rectangle $ABCD$ has area
$$24 \times 28 = 672 \text{ cm}^2.$$
On the front and back two triangles. Their total area is
$$2 \times \frac{1}{2} \times \text{base} \times \text{height}$$
$$= \text{base} \times \text{height}$$

= 24 × 32
= 768 cm².
The total area of the two sloping walls is
2 × 28 × 34.176
= 1913.86 cm².
(Check that you understand the 34.176...)
The sum of this is 672 + 768 + 1913.86
= 3354 8 cm².
Note that if you took the length of AW to be exactly 34 cm, the final answer would be 3344 cm².

Question 315

a. The volume of a pyramid is given by the equation
$$V = \frac{1}{3} \times \text{base} \times \text{height}$$
= $\frac{1}{3} \times 24 \times 28 \times 32$
= 7168 cm³.

b. Start by calculating the length of AC. You will need to consider a point, M, on the floor directly under Y.

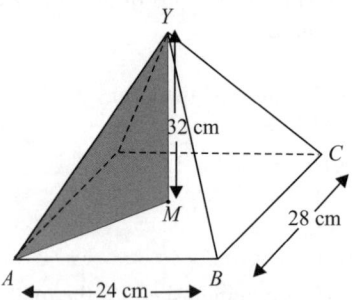

The diagonal of the base.
$AC^2 = 24^2 + 28^2$
∴ $AM = \frac{1}{2} \times \sqrt{24^2 + 28^2}$
= 18.4391 cm.
Now use Pythagoras' theorem with triangle AMY.
$AY^2 = 18.4391^2 + 32^2$
∴ $AY = \sqrt{18.4391^2 + 32^2}$
= 36.9324
≈ 37 cm, as required.

c. For the area of $\triangle YAB$, use Heron's formula as given to you in the exam;
$$A = \sqrt{s(s-a)(s-b)(s-c)}$$
Where, in this case,
$s = \frac{1}{2}(37 + 37 + 24)$
= 49
Thus,
$A = \sqrt{49(49-a)(49-b)(49-c)}$
= $\sqrt{49(49-37)(49-37)(49-24)}$
= $\sqrt{176\,400}$
= 420 cm³.

Question 316

a. The original height was 32 cm. Since 8 cm remains below, the amount that has been removed is 32 − 8 = 24 cm.
Thus the fraction that has been removed is
$\frac{24}{32} = \frac{3}{4}$.

b. The ratio of two lengths needs to be cubed to give the ratio of two volumes corresponding to those lengths.
Thus, our ratio is $\left(\frac{3}{4}\right)^3 = \frac{27}{64}$.

Solutions: C4

Question 317

a. $speed = \dfrac{distance}{time}$

$\therefore$ distance = speed × time
= 4 × 1.5 = 6 km

b. Here is the function $D_m = 4t$ sketched on their graph axes. Note that when $t = 5$, $D = 5 × 4 = 20$. Take special care to see that the graph starts at $t = 0$ and finishes at $t = 5$. Putting a small dot at each end of the graph makes it quite formal. I have removed some labels here for clarity.

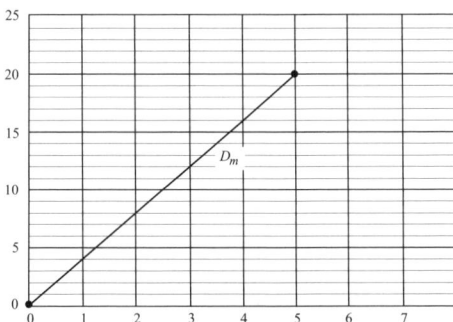

c. If Christos walks for 2 hours at a speed of 6 km/h, he will have walked 12 km. This is shown in the graph below (noting that it says 'on the same graph'...).

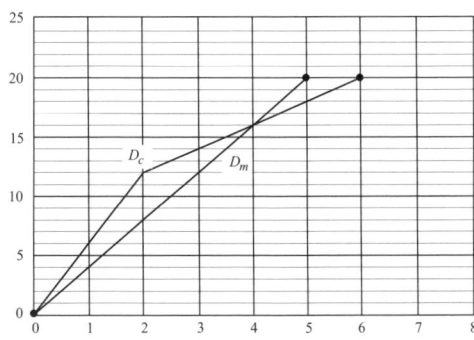

After that, however, he walks at a speed of 2 km/h. Now, he still has 8 km to go, so it will take him 4 more hours.
That brings the total time to 6 hours. The bend in the graph takes place when he changes speed.

d. Malinda eventually catches up to Christos at the time when the two graphs cross. According to the graph, this is at time $t = 4$ hours.

e. If
$$D_c = \begin{cases} at & 0 \leq t \leq 2 \text{ hours} \\ bt + h & 2 < t \leq d \text{ hours} \end{cases}$$
then $a = 6$ (his first speed), $b = 2$ (his second speed) and $d = 6$, the time when he reaches the destination, Snake Gully.
Finding h is a bit trickier.
Consider the equation $D_c = 2t + h$. We know that when $t = 2$, $D_c = 12$:
$\therefore \quad 12 = 2(2) + h$
$\therefore \quad h = 8$
Solution: $a = 6, b = 2, h = 8, d = 6$.

Question 318

a. i. The missing numbers are 4, 16 and 25.

ii. For clarity, some labels have been removed. Likewise, the coordinates of the points have been added here. For one mark, I am not sure that I would have added them during the exam...

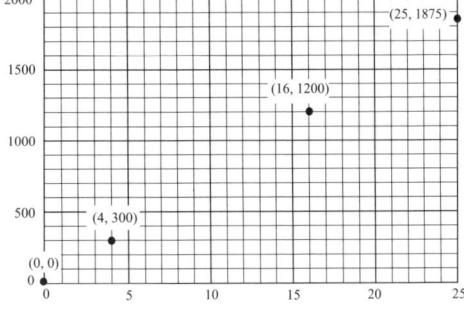

Solutions: C4

b. If you will pardon the temporary use of y and x, the equation of the straight line joining the dots is $y = mx + 0$ (the y-intercept is 0). The gradient is $\dfrac{1875}{25} = 75$.
Thus, using the correct symbols, $V = 75t^2$.
∴ $k = 75$.

c. During the first three hours (that is, $t = 3$ and $t^2 = 9$), Malinda drank $75 \times 9 = 675$ mL of water.

d.   i. Malinda started with 2000 mL. When she consumed 1000 mL,
$$V = 75 \times t^2$$
∴ $1000 = 75 \times t^2$
∴ $75 \times t^2 = 1000$
∴ $t^2 = \dfrac{1000}{75}$
∴ $t = 3.65148$
≈   3.65 hours

ii. At time $t = 3.65$ hours Malinda was at distance $4 \times 3.65 = 14.60$ km.
Christos was into the second phase of his walk, and uses equation $D_c = 2t + 8$
= $2(3.65) + 8$
= 15.30 km.
Thus, their distance apart is 0.7 km.
Comment. Keep a higher order of accuracy until the last line, *then* round off the answer.

Question 319

a. Given that $C = 8x + 2400$, and using the fact that $x = 400$,
$C = 8(400) + 2400$
= $3200 + 2400$
= 5600

b. Let $3000 = 8x + 2400$
∴ $3000 - 2400 = 8x$
∴ $8x = 600$
∴ $x = 75$ shirts.

c. Note that the graph of $C = 8x + 2400$ will have a vertical axis intercept of 2400. Mark it on the graph axes. Next, the end of the domain is where $x = 400$. Part **a**. shows us that the cost is $5600. Mark (400, 5600) on the axes:

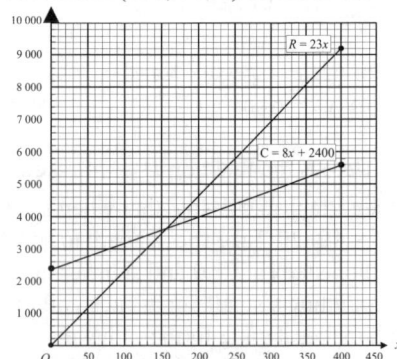

d. Look for where the lines cross on the graphs. The number of shirts looks closest to 160.

e. The profit is equal to the revenue minus the cost of making the shirts
$P = R - C$
= $23x - (8x + 2400)$
= $23x - 8x - 2400$
= $15x - 2400$
Watch that second '−'.

f. For 345 shirts, the profit is
$15x - 2400$
= $15(345) - 2400$
= $2775

Question 320

P = revenue − costs
∴ 3000 = revenue − 4800
∴   revenue = 7800
Now, this is for 250 jackets. That means the cost per jacket must be $\dfrac{7800}{250}$
= $31.20

Solutions: C4

Question 321

a. For 620 singlets, use the lower equation,
$$R_S = 6x + 2000$$
$$= 6(620) + 2000 = \$3720$$

b. The sketch graph for the hybrid equation
$$R_S = \begin{cases} 10x & x \leq 500 \\ 6x + 2000 & x > 500 \end{cases}$$
will be 'bent' at $x = 500$. Note that when $x = 500$, $R_s = 10 \times 500 = 5000$ because we will be using the upper equation. At the end of the scale, we have $x = 1000$. Using the second equation, we obtain $R_s = 6(1000) + 2000 = 8000$. The graph looks like this:

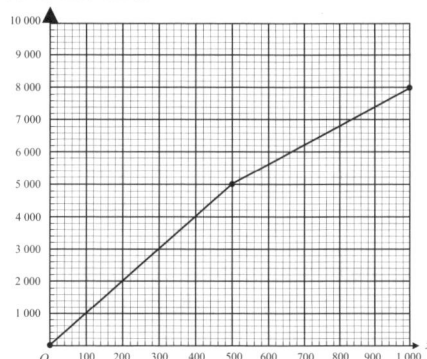

c. Just as before,
$$P = \text{revenue} - \text{costs}$$
$$\therefore \quad 2000 = \text{revenue} - (4x + 1500)$$
But this time, we seem to have a choice between equations of R_S. Let's try both. If we use $R_S = 10x$ then
$$2000 = 10x - (4x + 1500)$$
$$\therefore \quad 2000 = 10x - 4x - 1500$$
$$\therefore \quad 3500 = 6x$$
$$\therefore \quad x = 583$$
But this is a problem because we are only allowed to use the first equation if $x \leq 500$, so we should use the second equation.
$$2000 = 6x + 2000 - (4x + 1500)$$
$$\therefore \quad 2000 = 6x + 2000 - 4x - 1500$$
$$\therefore \quad 1500 = 2x$$
$$\therefore \quad x = 750 \text{ singlets}$$

Question 322

a. In effect we are being asked to identify which line has equation $20x + 15y = 900$. The y-intercept of that line is when $x = 0$
$$\therefore \quad 15y = 900$$
$$\therefore \quad y = \frac{900}{15}$$
$$= 60$$
It is graph A.

b. For a mere one mark, I will simply read the coordinates of the point of intersection from the graph. It looks awfully like (24, 28).

c. Since x relates to Xenon street and y relates to Yarra street,
Inequality 5: $x \leq 64$
Inequality 6: $y \leq 52$

d. You need to add the two new inequalities, $x \leq 64$ and $y \leq 52$, to the graph system – and then look for the answer region:

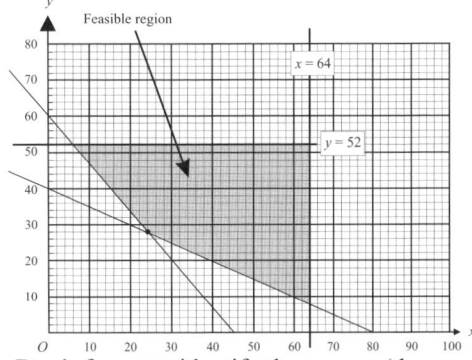

Don't forget to identify the answer (the feasible region).

e. Using the data given, the cost will be $C = 100x + 70y$.

f. Here, it is necessary to identify the corner points of the feasible region and to evaluate the cost at each point:

225

Solutions: C4

Corner	x	y	$C = 100x + 70y$
1	6	52	4240
2	64	52	10 040
3	64	52	7870
4	24	28	4360

The minimum cost is at the point with coordinates (6, 52). Answer: $4240.

g. The number of books is $2x + 3y$. This must *exceed* 150. The new inequality is $2x + 3y \geq 150$.

h. The new line has equation $2x + 3y = 150$. This line will have y-intercept given by $3y = 150 \therefore y = 50$ and x-intercept $2x = 150 \therefore x = 75$. Adding the new graph gives this:

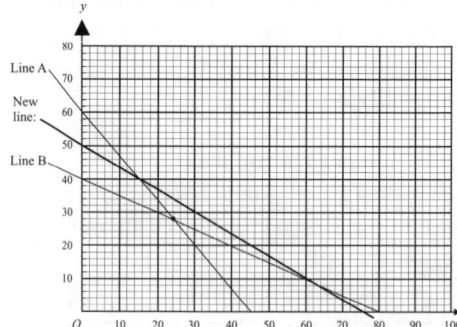

i. We are asked if the point (25, 32) is above, on or below the inequality $2x + 3y \geq 150$. Substitute (25, 32) into it:
$2x + 3y \geq 150$
$\therefore \quad 2(25) + 3(32) \geq 150$
$\therefore \quad 50 + 96 \geq 150$
$\therefore \quad 146 \geq 150$
Since this is a false statement, then (25, 32) does not obey the inequality. The answer is the word 'NO'.

j. i. The cost is increased so that the cost equation becomes
$C = 120x + 90y$
For 1 mark, substitute (25, 32) into the cost equation:
$C = 120(25) + 90(32)$

= $5400.

ii. The feasible region has now changed. The diagram now looks like this:

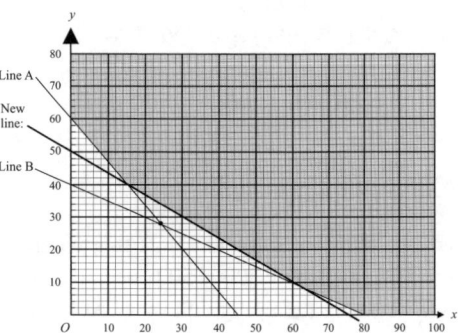

Once again we will need to check the costs at all of the corners. A new corner has been added. From the graph, it is (15, 40).

Corner	x	y	$C = 120x + 90y$
1	0	60	5400
2	80	0	9600
3	60	10	8100
4	15	40	5400

There are two points of lowest cost, (0, 60) and (15, 40). Therefore the cost will be minimum at all points on the line $20x + 15y = 900$ between $x = 0$ and $x = 15$.

Question 323

a. i. From the graph, a distance of 20 km gives a cost of $30.

ii. For a call-out fee of $10, the distances range from 'just above' 0 to 5 km. The answer is 5 km.

b. Note the open and closed circles carefully. The modified graph looks like this:

Solutions: C4

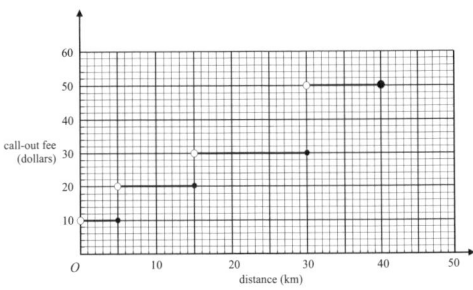

Question 324

a. For the moment assume an equation of the form $y = a + bx$. The y-intercept of this graph is 50 and the gradient is *negative*. To find the gradient, it is probably easiest to make a triangle as

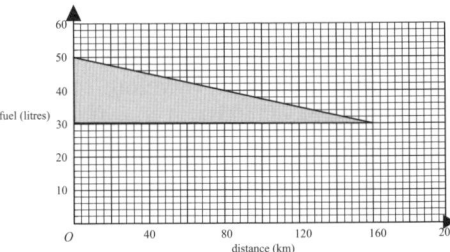

shown here:
The 'rise' of this triangle is 20 units are the run is 160 units. That means that
$\dfrac{\text{rise}}{\text{run}} = \dfrac{20}{160} = \dfrac{1}{8}$. The equation becomes
$y = 50 - \dfrac{1}{8}x$... or even better,

fuel $= 50 - \dfrac{1}{8}$ distance

b. For this line to 'hit the baseline', we require fuel = 0.

$\therefore \quad 0 = 50 - \dfrac{1}{8}$ distance

$\therefore \quad \dfrac{1}{8}$ distance $= 50$

$\therefore \quad$ distance $= 400$.

Compared to 160 km, this is an *additional* 240 km.

c. When he stops, there will be 12 litres remaining. Now, $3\dfrac{1}{2}$ litres for each of 18 minutes means $18 \times 3\dfrac{1}{2} = 63$ additional litres. The total will be $63 + 12 = 75$ litres.

Question 325

a. We are asked to add $20x + 25y = 200$ to the graph. The y-intercept of that graph is when $x = 0$. That is,
$\qquad 25y = 200$
$\therefore \quad y = 8$
The x-intercept is when $y = 0$. That is,
$\qquad 20x = 200$
$\therefore \quad x = 10$
The resulting graph:

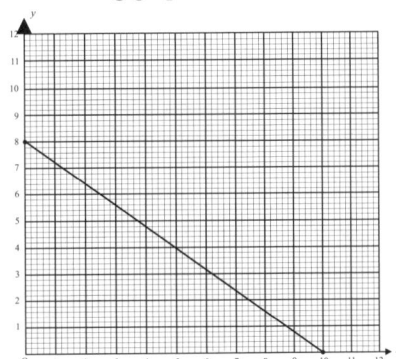

b. 'At least twice' leads to the result of $y \geq 2x$.

c. **i.** The required region is shaded:

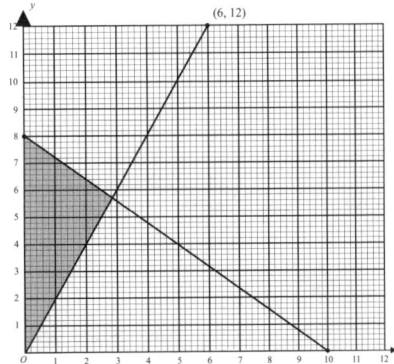

227

Solutions: C4

ii. The largest x-value inside the feasible region is about 2.8. But as 0.8 of a clip is somewhat inelegant, the largest practical value of x in the region is 2 clips.

You can find this by considering all of the points that have integer values. The points (2, 4), (2, 5) and (2, 6) all have $x = 2$.

d. $P = 40x + 30y$

e. i. This problem requires that whole numbers be used for x and y. The profit $\$P$ is defined by $P = 40x + 30y$ and the points should be chosen from the integer-valued points near the corners.

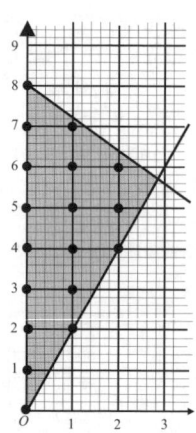

x	y	Profit $= 40x + 30y$
0	7	210
0	8	240
1	6	220
1	7	250
2	5	230
2	6	260

The largest profit ($260) is gained when there are 2 washes and 6 clips.

ii. Using the previous table, the maximum profit is $260.

Question 326

a. Note the sections on the graph that are flat. They represent the times when the family car has stopped.

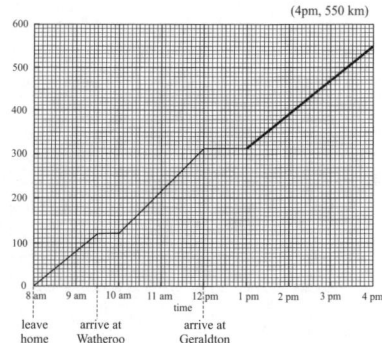

We are asked about their stop at Watheroo. On the graph, the flat Watheroo section at 9:30 am continues for exactly 30 minutes. That is how long the family stopped.

b. The end of the flat section at Watheroo is at time 10:00 am, where the distance is 120 km. The start of the 'at Geraldton' section is at 12:00 pm, where the distance is 310 km. Therefore, the distance travelled is 310 − 120 = 190 km.

c. The distance travelled in the section from Watheroo to Geraldton is 190 km, and the amount of time taken is 2 hours (see part b.). Therefore, the average speed is $\frac{190}{2} = 95$ km/h.

d. Here, we are being asked about the 30 section of the journey from Geraldton (which starts at 1 pm) to Hamelyn. The journey takes three hours, so it will finish at 4 p.m. the distance travelled is found by using the equation

$$\text{speed} = \frac{\text{distance}}{\text{time}}$$

$$\therefore \text{ distance} = \text{speed} \times \text{time}$$
$$= 80 \times 3$$
$$= 240 \text{ km.}$$

Add 240 km to the distance at Geraldton (310 km), and you get 550 km. Now join the coordinate (1, 310) with (4, 550) on the graph.

228

Solutions: C4

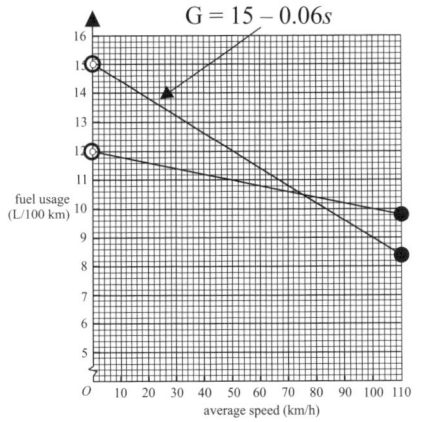

On my graph, the journey from Geraldton to Hamelin is in shown in bold.

Question 327

a. For a mere one mark, just read the answer off the graph. When the speed is 60, the fuel consumption is 10.8 L/100 km. For those who prefer to use algebra, compute $P = 12 - 0.02(60)$.

b. To draw the graph of $G = 15 - 0.06s$ it is best to create two dots and join them.
When $s = 0$, $G = 15 - 0 = 15$.
When $s = 110$, $G = 15 - (0.06)110 = 8.4$.
Thus, the two dots $(0, 15)$ and $(110, 8.4)$ are plotted and joined. Note that you are required to create a graph 'up to 110 km/h'.

c. For one mark, the question is asking us to identify when the 'gas graph is lower than the petrol graph'.
First, where do they intersect?
$12 - 0.02s = 15 - 0.06s$
∴ $0.06s - 0.02s = 15 - 12$
∴ $0.04s = 3$
∴ $s = 75$
Second, identify the correct region on the graph, taking note that it does not say 'less than or equal to'.
Answer: Fuel usage of gas will be less than fuel usage of petrol when the speed is greater than 75 km/h.

d. The gas consumption will be
$G = 15 - 0.06s$
$= 15 - 0.06(85)$
$= 9.9$ L/100km
Thus, for a trip of 100 km, the cost is
9.9×0.80
$= 7.92$
$= 7$ dollars and 92 cents.
Read the question again. Be sure to identify the units that are required here.

Question 328

a. Inequality 5 means that the total driving time (both petrol and gas) must not exceed 24 hours.

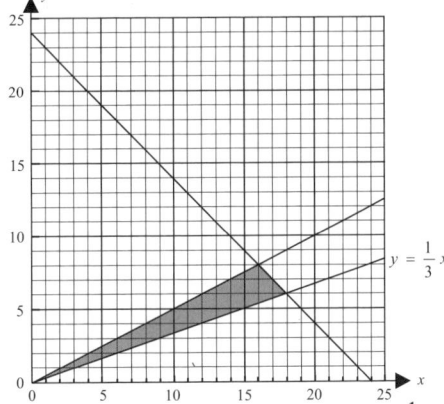

b. I have added in the graph of $y = \frac{1}{3}x$ and I have shaded the feasible region on the diagram. Note that the point $(24, 8)$ lies on the line $y = \frac{1}{3}x$.

c. Locate the point on the graph where $G = 10$ and $P = 5$ (note that $10 + 5 = 15$). I have marked it with an 'X'. That point is in the feasible region, therefore, the

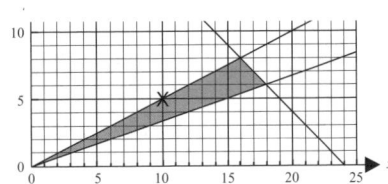

229

Solutions: C4

Goldsmiths will comply with all the constraints.

d. The answers here must lie on the line $x + y = 24$ because they plan to drive for 24 hours. The answers must also be in the feasible region, which by now has become a single line section.

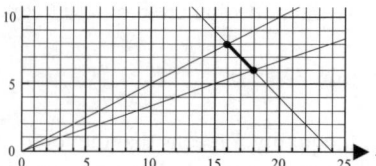

From the graph, the maximum and minimum 'G' values are 18 and 16 hours respectively.

Solutions: C5

Question 329

a. The value of the car follows this pattern:

Time	Value
0	40 000
1	40 000 × 0.75 = 30 000
2	30 000 × 0.75 = 22 500
3	22 500 × 0.75 = $16 875

So:
i. After one year, the value is $30 000.
ii. After three years, the value is $16 875.

b. The value of the computer equipment after two years is $40 000 – 2 × 8000 = $24 000.

c. Here is a table showing the values of the car and of the computer:

Year	Car value	Computer value
0	40 000	40 000
1	30 000	32 000
2	22 500	24 000
3	16 875	16 000

Thus, after 3 full years, the value of the car first exceeds the value of the computer.

d. Each year, $8000 is paid in interest. This is compared with the original value of $40 000. Thus, the flat % rate is:

$$\frac{8000}{40\,000} \times 100 = 20\%$$

Question 330

a. The total amount of Briony's repayments to the building society will be 1500 × 12 × 10 = $180 000.

b. The total interest is 180 000 – 70 000 = $110 000.

c. The $110 000 is spread over 10 years. So it is $11 000 per year. Thus, the flat rate of interest (%) is:

$$\frac{11000}{70000} \times 100 = 15.7\%$$

Question 331

a. $80\,000 + \dfrac{8.5}{100} \times 80\,000 - 10\,000$
$= \$76\,800$

b. **i.** $Q = \$10\,000$, $R = 1.085$

ii. Using $A = PR^n - \dfrac{Q(R^n - 1)}{R - 1}$

with $P = 80\,000$, $n = 10$
∴ $A = \$32\,527.68$ (nearest cent)
If you wish to use a graphical calculator, make special note of the + and – signs:
N = 10
I% = 8.5
PV = 80 000
PMT = –10 000
FV = ?
P/Y = 1
C/Y = 1
Once again, FV = $32 527.68.

c. Using the graphical calculator to note that the future value is $80 000, and choosing any very large value for N,
N = 100
I% = 8.5
PV = 80 000
PMT = ?
FV = -80000
P/Y = 1
C/Y = 1
Then, the payments are seen to be $6800.00.

Solutions: C5

Question 332

a. **i.** On 2 February 2002, the balance dropped from $3600 to $3000. Therefore, $600 was withdrawn.
ii. On 14 May 2002, the balance of $3000 was increased by $85.50, giving a new balance of $3085.50.

b. For the month of December 2001, the minimum balance was $2400. This earned interest of :

$$2400 \times \frac{0.3}{100} \times 1 = \$7.20$$

Question 333

a. $I = \frac{Prt}{100}$

$\therefore R = \frac{100(416)}{(4000)2}$

$= 5.2\%$ pa

b. Sally will earn compound interest according to the formula:
$P' = P \times R^n$ where $R = 1 + \frac{4.8}{100} = 1.0048$
$= 4000 \times 1.048^2$

Now, comparing this with Sally's formula total amount $= 4000 \times c \times c$, it means that her value of c should be 1.048.

c. Compound interest:

$P' = P \times R^n$
$\Rightarrow 4416 = 4000 \times R^2$
$\Rightarrow R^2 = \frac{4416}{4000}$
$= 1.104$
$\therefore R = \sqrt{1.104}$
$= 1.050714$

Correct to two decimal places, this means an annual interest rate of 5.07%.

Question 334

a. Her car lost $23\,600 - 7000 = 16\,600$ over five years. That is a flat rate of:

$$\frac{16\,600}{23\,600} \times \frac{100}{5}$$
$= 14.43\%$

b. Method 1: Use a graphical calculator with the 'annuities' formula. Set:
N = 5
I% = ?
PV = –23 000
PMT = 0
FV = 7000
P/Y = 12

Solving for I% gives 21.6% (correct to 1 decimal place.)

Method 2: Use algebra:

Let $A = Pr^n$
$= P \times (1 - \frac{r}{100})^n$

$\therefore 7000 = 23\,600 \times (1 - \frac{r}{100})^5$

$\therefore \frac{7000}{23\,600} = (1 - \frac{r}{100})^5$

$\therefore 1 - \frac{r}{100} = \sqrt[5]{\frac{7000}{23\,600}}$

$\therefore 1 - \frac{r}{100} = 0.7842$

$\therefore \frac{r}{100} = 0.21576$

$\therefore r \approx 21.6\%$

Question 335

a. $A = PR^n - \frac{Q(R^n - 1)}{R - 1}$

i. $R = 1 + \frac{8/4}{100} = 1.02$

ii. $n = 4 \times 4 = 16$

Solutions: C5

b. Using a calculator with the above data and $P = 20\,000$ and $Q = 1500$, I found that $A = -\$503.21$. You may interpret the negative sign as meaning that the loan has been (more than) fully paid.

If you have time to spare, you might like to also calculate that the loan is exactly paid in 15.6 quarters.

Question 336

Brad wants to buy a coffee machine for his café. Crazy Bill's normally sells them for $3450, but they have a special discounted price of $3100 for this week.

a. Percentage discount is found using
$$\frac{\text{drop in price}}{\text{original price}} \times \frac{100}{1}$$
$$= \frac{2400 - 3100}{3450} \times \frac{100}{1}$$
$$= 10.1449$$
$$\approx 10.1\%$$

b. **i.** The total cost of the machine is
$$200 + 275(12)$$
$$= \$3500$$

ii. Since the original cost was $3100 less the $200, $400 interest was charged. This represents $\frac{400}{2900} \times 100$
$$= 13.7931$$
$$\approx 13.8\%$$

c. **i.** $n = 24$ (monthly over two years), $P = 3100$ and $A = 0$ (to be exactly paid off.)

ii. Using the graphical calculator, set these values:
N = 24
I% = 9
PV = –3100
PMT = (anything or blank)
FV = 0
P/Y = 12
C/Y = 12

Solving for PMT gives $141.62 (correct to 2 decimal places). But that rounding down causes a problem. At the end of the 24 payments, 7 cents will still be owing. The bank won't be at all pleased. So it is safer to pay back $141.63 each time.

iii. The total cost will be 24 multiplied by this monthly payment:
24×141.62
$= 3398.88$
$\approx \$3399$ (correct to the nearest dollar)

d. Crazy Bill's deal means that Brad pays $3500. Discount King's cost will be $3399. Therefore, Discount King offers the best deal by an amount of $101.

Question 337

a. The value this year is 0.85 multiplied by last year's value (15% depreciation).

Year	Value
0	$3100
1	$2635
2	$2239.75
3	$1903.79

The depreciated value of the machine after three years is $1904 (correct to the nearest dollar).

b. 15 000 cups of coffee multiplied by $0.03 is $450 per year. After three years this is $3 \times 450 = \$1350$. The machine will be worth $3100 - 1350 = \$1750$.

c. If we call the unit cost per cup of coffee 'c', which seems a good choice...

$3100 - 15\,000 \times c \times 3 = 1904$
$\therefore\ 3100 - 1904 = 15\,000 \times c \times 3$
$\therefore\ 15\,000 \times c \times 3 = 3100 - 1904$
$\therefore\ 45\,000 \times c = 1196$
$\therefore\ c = \dfrac{1196}{45\,000}$
$= 0.02657$ dollars
≈ 2.7 cents per cup.

Solutions: C5

Question 338

a. i. Remy makes 6 payments of $120 which means he pays $720.

ii. Since he paid $720 but the camera cost $650, he has paid $70 in interest.

b. i. $I = \dfrac{Pr\,t}{100}$

$\therefore\ 70 = \dfrac{650 \times r \times 0.5}{100}$

(Note that we want an *annual* interest rate).

$\therefore\ 7000 = 650 \times r \times 0.5$

$\therefore\ r = \dfrac{7000}{650 \times 0.5}$

$= 21.538$

$\approx 21.5\%$ (correct to one decimal place).

ii. As a loan is paid off, the interest, in dollars, should reduce. Simple interest rate does not take this into account. Effective interest is larger because it allows for this.

c. 10% is calculated by multiplying by 1.1. So,

$1.1 \times \text{price} = 650$

$\therefore\ \text{price} = \dfrac{650}{1.1}$

$= 590.909$

$\approx \$590.91$

d. Flat rate of 12%:

$I = \dfrac{P \times r \times t}{100}$

$= \dfrac{650 \times 12 \times 5}{100}$

$= \$390$

That means the camera dropped in value by $390, making its new value
650 - 390 = $260

15% reducing value means that Remy must multiply the current value by 0.85 to find the next value. So after 5 years, the value is

650×0.85^5

$= 288.4084$

$\approx \$288.41$

That means that the flat rate balance method gave a smaller final value. The difference is 288.41 − 260 = $28.81.

Question 339

a. From the information given
$A = 0$ (which represents full repayment)
$n = 12 \times 4 = 48$
$P = 12\,000$

b. Using the TVM solver routine on my calculator, I put these values in:
$N = 48$
$I\% = 7.5$
$PV = -12\,000$
$PMT = ?$
$FV = 0$
$P/Y = 12$
$C/Y = 12$
The payment came to $290.15.

c. If Anna made 48 payments of $290.15 then she would have paid $13\,927.20. The amount borrowed was $12\,000, so the interest was $1927.20. To the nearest dollar, this is $1927.

d. Using the TVM solver routine on my calculator, I put these values in:
$N = 6$
$I\% = 7.5$
$PV = -12\,000$
$PMT = 290.15$
$FV = ?$
$P/Y = 12$
$C/Y = 12$
The future value comes to $10\,688.78. But this is the amount still owing, not what was paid off.
That value is $12\,000 - \$10\,688.78$ which is equal to $1311.22, or to the nearest dollar, $1311.

234

Solutions: C5

e. Treat this question as a new loan, borrowing $10 688.78 and with an unknown 'n' value. Note that there are only 42 payments to be made:
$N = 42$
$I\% = 8$
$PV = -10\,688.78$
$PMT = ?$
$FV = 0$
$P/Y = 12$
$C/Y = 12$

i. So, $n = 42$ and $P = \$10\,688.78$.

ii. Using TVM solver again, this gives the monthly payment as $293 (to the nearest dollar).

Question 340

a. The computer cost $4000 and he paid $4560. Therefore he paid $560 in interest.

b. The flat rate is given by
$$I = \frac{P \times r \times t}{100}$$
$$\therefore 100 \times I = P \times r \times t$$
$$\therefore \frac{100 \times I}{P \times t} = r$$
$$\therefore r = \frac{100 \times I}{P \times t}$$
$$= \frac{100 \times 560}{3500 \times 2}$$
$$= 8\%$$

c. The effective rate is given by
$$\frac{2n}{n+1} \times \text{flat rate}$$
$$= \frac{2 \times 24}{24+1} \times 8$$
$$= 15.36$$
$$\approx 15.4\% \text{ correct to one decimal place.}$$

d. Simple interest rate does not take into account the reduction of the principal as each payment is made.

Question 341

a. Since the total lost is $3000 over 5 years, the depreciation is $600 per year. Compared to the original $4000, this is
$$\frac{600}{4000} \times 100$$
$$= 15\%$$

b. Reducing balance means that each year, you must multiply by $1 - \frac{r}{100}$. At the end of 5 years, the value would be $4000 \times \left(1 - \frac{r}{100}\right)^5$. But we know this is equal to 1000:
$$\therefore 4000 \times \left(1 - \frac{r}{100}\right)^5 = 1000$$
$$\therefore \left(1 - \frac{r}{100}\right)^5 = 0.25$$

Take the 5th root of both sides:
$$1 - \frac{r}{100} = 0.25^{1/5}$$
$$\therefore 1 - 0.25^{1/5} = \frac{r}{100}$$
$$\therefore r = 100(1 - 0.25^{1/5})$$
$$= 24.21417$$
$$\approx 24.2\%$$

Question 342

a. i. The three values required are:
* $P = 10\,000$
* $R = 1 + \frac{4.8}{12} = 1.004$
* $n = 60$.

ii. At the end of 5 years, $n = 5 \times 12 = 60$ and $10\,000 \times (1.004)^{60} = 12\,706.41$.
Answer: $12 706.41.

b. It is definitely best to use TVM solver:
$N = 60$
$I = 4.8$

Solutions: C5

PV = 4000
PMT = 100
FV = ?
P/Y = 12
C/Y = 12
Putting the cursor on FV and pressing SOLVE provides an answer of -11848.58.
Answer: $11 848.58.

c. i. Once again, it is definitely best to use TVM solver, noting that it is for 3 years.
N = 36
I = 4.8
PV = 6915.90
PMT = ?
FV = -13 000
P/Y = 12
C/Y = 12
Putting the cursor on PMT and pressing SOLVE provides an answer of 128.80.
Answer: $129.80.

ii. In the first plan $4000 is given as deposit. Then 24 × $100 payments are made. Then 36 payments of $129.80 are made. This totals $11 072.80.
But Option C gives back $13 000. This means that the interest is
13 000 − 11 072.80 = $1927.20.

Question 343

a. i. The machine will depreciate annually by 10% of $60 000 = $6000.

ii. After three years, the machine will reduce in value by 3 × 6000 = 18 000. That means its value will be 60 000 − 18 000
= $42 000.

iii. Let n be the unknown number of years.
∴   60 000 − 6000n = 12 000
∴   48 000 = 6000n
∴ n = 8 years.

b. i. Multiplying by 0.85 (as we can see 0.85 in the equation) reduces an amount by 15%.

ii. The value of the machine after three years will be
$V = 60\,000 \times (0.85)^n$
$= V = 60\,000 \times (0.85)^3$
$= \$36\,487.50$

iii. Start by letting the value equal $12 000:
$12\,000 = 60\,000 \times (0.85)^n$
∴ $(0.85)^n = \dfrac{1}{5}$.

Now, either use logs, a solving program or a table of values. I will use logs:
$(0.85)^n = \dfrac{1}{5}$
∴ $\log(0.85)^n = \log \dfrac{1}{5}$
∴ $n \times \log(0.85) = \log \dfrac{1}{5}$
∴ $n = \log \dfrac{1}{5} \div \log(0.85)$
$= 9.9031$
As the value is falling, the value will be below $12 000 during the 10th year.

c. Equate the two statements:
$60\,000 \times (0.85)^n = 60\,000 - 6000n$
I chose to sketch two graphs. With a window −1 to 10 for x and −10 to 60 000 for y, I found the intersection to be at $x = 6.5522$. The graph tells you to round UP. That is, the 7th year.

Question 344

The year by year prices will form a geometric progression with a = 60 000 and r = 1.02. After 8 years, this will be
$60\,000 \times 1.02^8$
$= 70\,299.56$
$\approx \$70\,300$.

Solutions: C5

Question 345

a. Simple interest is defined by
$$I = \frac{P \times r \times t}{100}$$
$$= \frac{7000 \times 6.25 \times 8}{100}$$
$$= 3500$$
But this must be added to the invested amount of $7000
$$= \$10\,500$$

b. 6% per annum compounding quarterly for eight years means 32 interest periods with a quarterly rate of $\frac{6}{4}$%
The value will be
$$10\,000 \times (1 + \frac{6}{100 \times 4})^{32}$$
$$= 16\,103.24$$
$$\approx \$16\,103$$

c. Use the annuities function on your graphical calculator. The time period here is the month. There are 96 months, and since money is being 'put in', -200 is the payment amount:
$N = 96$
$I\% = 6.5$
$PV = -500$
$PMT = -200$
$FV = ?$
$P/Y = 12$
This produces an FV value of $25 935.39 or approximately $25 935.

Question 346

First find the amount of the monthly payment. Use the annuities function on your graphical calculator. The time period here is the month. There are 24 monthly payments of an unknown amount and in order to be 'paid off', the future value will be $0.
$N = 24$
$I\% = 10$
$PV = -20\,000$
$PMT = ?$
$FV = 0$
$P/Y = 12$
This produces a PMT value of $922.90.
The total amount to be paid will be
$$24 \times 922.90$$
$$= \$22\,149.60$$
Since the item costs $20 000, the interest paid will be $2149.60 or, to the nearest dollar, $2150.

Question 347

a. i. 25% deposit means
$$\frac{25}{100} \times 7000$$
$$= \$1750$$

ii. The balance due it is
$$7000 - 1750$$
$$= \$5250.$$
Splitting this over 24 payments means
$$\frac{5250}{24}$$
$$= 218.75$$
$$= 218 \text{ dollars and } 75 \text{ cents.}$$
Note the units that are specified in the question.

b. i. The amount paid is equal to
$$500 + 36 \times 220$$
$$= \$8420$$

ii. First, calculate the amount of interest in dollars.
$$8420 - 7000$$
$$= \$1420$$
Next, take the deposit into account.
$P = 7000 - 500 = 6500$
Now use the formula
$$I = \frac{P \times r \times t}{100}$$
$$\therefore \quad 1420 = \frac{6500 \times r \times 3}{100}$$
$$\therefore \quad 1420 = 195r$$
$$\therefore \quad r = 7.282 \approx 7.3\%$$

Solutions: C5

c. The discount is $7000 \times \dfrac{15}{100} = 1050$, so that the price becomes
$7000 - 1050 = \$5950.$

Question 348

a. The amount of interest he will pay each month is
$$30\,000 \times \frac{9}{100} \times \frac{1}{12}$$
$= \$225.$

b. I have chosen to use the finance screens of my graphical calculator with the following values:
- N = 60
- I% = 9
- PV = −30 000
- PMT = 400
- FV = ?
- P/Y = 12
- C/Y = 12

Asking for the future value, FV, gives
16 800.776
≈ $16 801

c. Using the finance screens of my graphical calculator again, I entered the following values:
- N = 60
- I% = 9
- PV = −30 000
- PMT = 400
- FV = 0
- P/Y = 12
- C/Y = 12

Asking for the payment value, PMT, gives
622.7507
≈ $622.75

Question 349

a. Let x be the price before the GST was added.
$x \times 1.10 = 900$

$\therefore \quad x = \dfrac{900}{1.10}$

$ = \$818.18$

b. i. Over a 5 year period, the fax machine dropped in value from $900 down to $300, a loss of $600. The annual depreciation is
$$\frac{600}{5} = \$120.$$

ii. If Khan sends five faxes a year for five years, then he sends 1250 faxes. The depreciation will be 1250×0.46. thus, the value of the fax machine will become
$900 - 1250 \times 0.46$
$= \$325.$

Question 350

The books in Khan's office are valued at $10 000.

a. The new value will be
$$10\,000 \times \left(1 - \frac{12}{100}\right)^5$$
$= 5277.3192$
$\approx \$5277.$

b. We require that
$$4000 = 10\,000 \times \left(1 - \frac{r}{100}\right)^5$$

$\therefore \quad \left(1 - \dfrac{r}{100}\right)^5 = 0.4$

$\therefore \quad 1 - \dfrac{r}{100} = \sqrt[5]{0.4}$

$\therefore \quad 1 - \dfrac{r}{100} = 0.832553$

($\sqrt[5]{0.4}$ was obtained by pressing 0.4^(1.5) including those brackets...)

$\therefore \quad \dfrac{r}{100} = 0.167447$

$\therefore \quad r = 0.167447$

$\approx 16.7\%$ pa.

Solutions: C6

Question 351

a. i. I found the shortest path to be this:

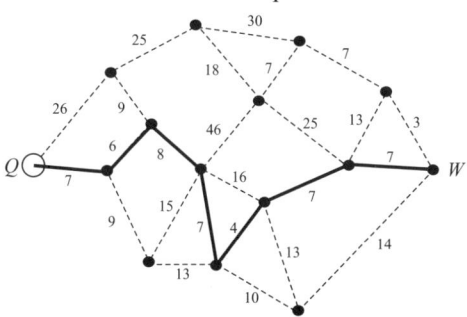

ii. The length of the shortest path is
$7 + 6 + 8 + 7 + 4 + 7 + 7 = 46$ km.

b. i. If he returns to the start then it is a circuit. If all edges must be visited once, then it is a Hamiltonian circuit.

ii. Here is a Hamiltonian circuit:

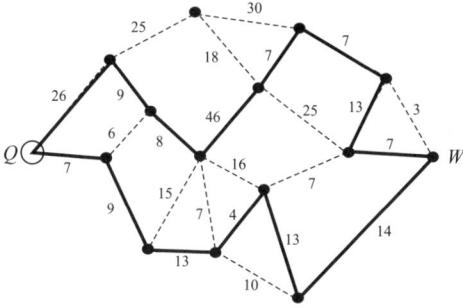

Question 352

a. The completed table will have 1 hour at the end of the A line, 4 hours in the D line, 19 hours on the F line and 12 hours on the K line.

b. The critical path on a diagram looks like this:

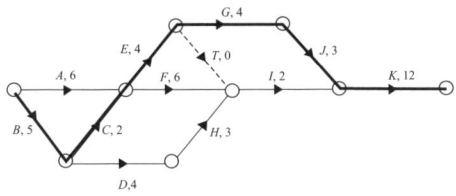

It is path B–C–E–G–J–K.

c. On the modified project, the critical path is B–C–F–I and the time taken is $5 + 2 + 6 + 2 = 15$ hours.

d. i. In order to maintain the critical path, B can be reduced by 1 unit and F can be reduced by 2 units. Reduce them any more and the critical path would change to include A or E/T. Answer: B and F.

ii. From part **i.**, the reduction is $1 + 2 = 3$ hours.

iii. The cost will be $1 \times 100 + 2 \times 50 = \$200$.

Question 353

a. The degree of vertex W is 5.

b. It is fastest to go via C both on the way to D as well as on the way back to W. The distances will be:
$6 + 6 + 6 + 12 = 30$ km.

c. i. Since every vertex is to be reached, a Hamiltonian path is required. Note that a circuit is not asked for.

ii. I found two paths. One of them is shown here:

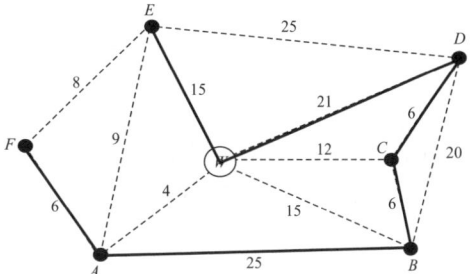

The other is *E-W-C-D-B-A-F*.

d. i. If all of the vertices are of even order, then an Euler circuit is possible. This is not the case here.

ii. The only path I could find was *W-E-D-W-A-F-E-A-B-W-C-D-B-C*. It finishes at *C*. It is a bit difficult to illustrate this without being messy. Try this:

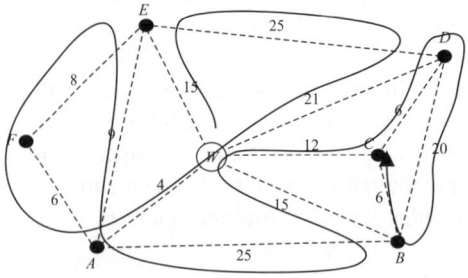

Question 354

a. Before activity *H* can start, all three of activities *D*, *F* and *G* must have finished. The longest path up until the start of activity *H* is *B-E-G* which takes 50 minutes.

b. The critical path of this circuit is *B-E-G-H-K*. Thus, it takes 110 minutes before activity *H* can start. *H* takes 15 minutes. Total time: 125 minutes.

c. i The critical path for this network is *B-E-G-H-K*. Since activity *F* is not on the path, it may be increased without delaying this repair. To be a little more precise, the slack time for *F* is 12 minutes. See part **ii.** below.

ii. Path *A-C-F* is of duration 38 minutes. Path *B-E-G* is of duration 50 minutes. Thus, there are 12 extra minutes available for activity *F*. This means that it can be increased to 32 minutes without delaying the repair.

d. i. The diagram should include a connection between *G* and the start of *J*. This 'dummy activity' (I have called it *L*) has duration 0 minutes and should be fully labelled.

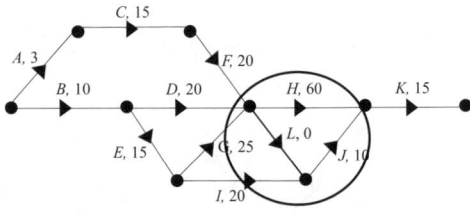

ii. Path *A-C-F* is of duration 58 minutes. Path *B-E-G* (on the critical path) takes 50 minutes. Thus *H* can start at 58 minutes at the latest. Note that there is also an argument to say that it should be 63 minutes by comparing path *B-E-G-L-J* of duration 115 minutes with *A-C-F-H* of duration 118 minutes. Three extra minutes are required.

iii. Path *A-C-F-H-K* takes $118 + 15 = 133$ minutes.

iv. You can delay activity for any of *B,E,G* or *J* by 3 minutes. Activity *I* can be delayed by $5 + 3 = 8$ minutes. But *D* can be delayed by $50 - 30 = 20$ minutes. So, here *D* is the winner.

Question 355

a. Since Harriet will play the drums, George can't play the drums. That means George must play the guitar.

b.

Person	Position
Harriet	Drums
Ian	Saxophone
Keith	Keyboard

Solutions: C6

Question 356

a. This is because it does not make sense for a musician to compete against themself.
b. The matrix indicates that Josie should have '1 outgoing line'. It is shown here:

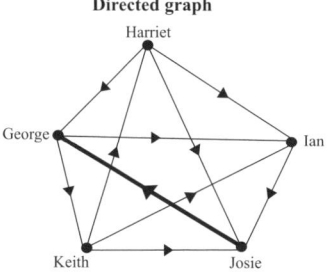

c. From the directed graph, it can be seen that George beats Keith but Keith beats Ian. This is a classic two-step dominance.

d. From the directed graph, it can be seen that Keith beats Harriet who beats Josie. Also, Keith beats Ian who also beats Josie. So there are two two-step dominances. Thus, $x = 2$.

e. Start by adding together the two matrices that we have been given.
Matrix 1 + Matrix 2 is

$$\begin{array}{c c} & \begin{array}{c c c c c} G & H & I & J & K \end{array} \\ \begin{array}{c} G \\ H \\ I \\ J \\ K \end{array} & \left[\begin{array}{c c c c c} 0 & 1 & 2 & 2 & 1 \\ 2 & 0 & 2 & 2 & 1 \\ 1 & 0 & 0 & 1 & 0 \\ 1 & 0 & 1 & 0 & 1 \\ 2 & 1 & 2 & 3 & 0 \end{array} \right] \end{array}$$

Now add the rows. The result is the following set of dominances:

$$\begin{array}{c} G \\ H \\ I \\ J \\ K \end{array} \left[\begin{array}{c} 6 \\ 7 \\ 2 \\ 3 \\ 8 \end{array} \right]$$

Thus Keith is first and Ian is last.

Question 357

a. Start by noting how G and H must finish and join at the start of I. The result is shown here:

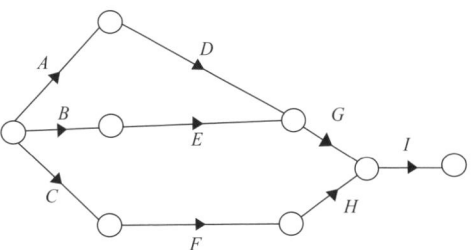

b. There are three complete paths. They are $ADGI$, $BEGI$ and $CFHI$. One of these is the critical path and so it will create a list of 4 critical activities. Since there are 9 activities altogether, there must be 5 that are non-critical.

c. From the data given, Activities A and C have some variation in possible start time (float time $\neq 0$) so they are not critical. This leaves Activity I. The only other path with activity I is $BEGI$.

d. The start time for activity I is 12 hours leaving $19 - 12 = 7$ hours for activity I.

e. Start by identifying the 1-hour float time for Activity C. We are told that its duration is 3 hours. We are seeking the maximum value of the combined duration of activities F and H. It must be true that Activity I started at time 12 hours, less the 3 hours for C and less the 1 hour float time for C. This tells us the maximum combined duration is 8 hours.

Question 358

a. The minimum is 4 edges.

241

Solutions: C6

b. There are several different possible answers. Here is one of them:

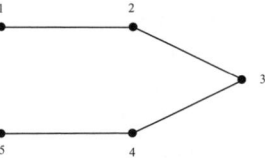

Question 359

a. Considering the degrees of the vertices from *A* to *G* in that order.
$4 + 2 + 5 + 2 + 4 + 4 + 3 = 24$

b. i. I started at the odd order vertex *G*.

ii. If you can manage to walk on each road exactly once, the total distance will be 2800 m. The path I took was *GAFG EDCF ECBA C*.

c. It is important to note that if Michelle starts at *F*, that she must visit each vertex exactly once and return to *F*.
Now, starting in the centre, it is hard to avoid using one of the roads twice in order to return to *F*. Since 150 is the shortest path, I have chosen to use it twice as follows: *FCBA GEDC F*.

Question 360

a. The capacity of Cut 1 is $14 + 8 + 13 + 8 = 43$. Note: ignore the 10.

b. The maximum number of seats will be determined by the capacity of a second cut shown here:

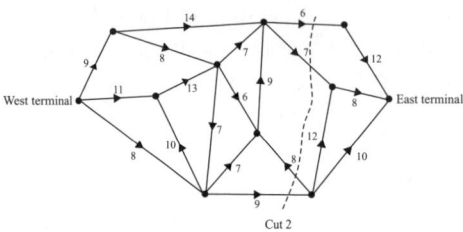

The maximum number of seats therefore, is $6 + 7 + 9 = 22$ seats.

c. From the very start, there are restrictions. Since there are 10 children, the train with a capacity of 11 must be used. Following that, trains with seating capacities of 13, 7,7 and 8 must be chosen. Therefore, the maximum number of children is 7.

Question 361

a. The critical part of this project is *B–C–F–H–I*. This part has 19 weeks duration.

b. The path *ADG* takes eight weeks. The critical path, finishing up to the start of *I*, takes $4 + 3 + 4 + 2 = 13$ weeks. Therefore, there are five weeks which can be allocated to slack time along the path *ADG*. This includes activity *D*.

c. Watch out for the word **not** in this question. We have a choice of activities here. Reducing *A*, *G* or *E* does not reduce the critical path, whereas reducing *C* or *F* would. Answer: *A*, *G* or *E*.

d. Reducing activities *C* and *F* by 2 weeks each would put activity *E* on the critical path because $C + F$ would only be three weeks duration. Therefore, reduce *E* by 1 unit as well. The critical path is now of 15 weeks duration.

e. The additional costs of the $2 + 2 + 1$ weeks of reduction will be 5×5000 which works out to be $25 000.

Solutions: C7

Question 362

a.

The order of $\begin{bmatrix} 2500 & 3400 & 1890 \\ 1765 & 4588 & 2456 \end{bmatrix}$ is 2×3.

b. **i.** $M = QP$

$= \begin{bmatrix} 2500 & 3400 & 1890 \\ 1765 & 4588 & 2456 \end{bmatrix} \begin{bmatrix} 14.50 \\ 21.60 \\ 19.20 \end{bmatrix}$

$= 2 \times 3 \quad \times \quad 3 \times 1$

Expect a 2×1 matrix answer, keeping the 2 decimal places of accuracy:

$= \begin{bmatrix} 145\,978.00 \\ 171\,848.50 \end{bmatrix}$

ii. The number of units of each multiplied by the selling price gives the total revenue for each of the products A, B and C at Eastown and Noxland.

c. Matrix P has order 3×1. Matrix Q has order 2×3. The multiplication PQ is not possible because the number of columns of P is not equal to the number of rows of Q. 'The orders are not conformable for multiplication.'

Question 363

a. The percentages must be changed into decimal values for the transition matrix. The resulting matrix looks like this:

$T = \begin{bmatrix} 0.80 & 0.09 & 0.10 \\ 0.12 & 0.76 & 0.05 \\ 0.08 & 0.15 & 0.85 \end{bmatrix}$

b. The matrix K_0 looks like this:

$K_0 = \begin{bmatrix} 300\,000 \\ 120\,000 \\ 180\,000 \end{bmatrix}$

c. The product $T \times K_0$ is

$\begin{bmatrix} 0.80 & 0.09 & 0.10 \\ 0.12 & 0.76 & 0.05 \\ 0.08 & 0.15 & 0.85 \end{bmatrix} \times \begin{bmatrix} 300\,000 \\ 120\,000 \\ 180\,000 \end{bmatrix}$

$= \begin{bmatrix} 268\,800 \\ 136\,200 \\ 195\,000 \end{bmatrix}$

d. What is being sought here is the product of $T^{\text{large number}} \times K_0$ and then $T^{\text{larger number}} \times K_0$ to show that, within acceptable limits, they are the same. I tried the power of 40:

$\begin{bmatrix} 0.80 & 0.09 & 0.10 \\ 0.12 & 0.76 & 0.05 \\ 0.08 & 0.15 & 0.85 \end{bmatrix}^{40} \times \begin{bmatrix} 300\,000 \\ 120\,000 \\ 180\,000 \end{bmatrix}$

$= \begin{bmatrix} 194\,982.90 \\ 150\,513.35 \\ 254\,503.75 \end{bmatrix}$

Now try the power of 45:

$\begin{bmatrix} 0.80 & 0.09 & 0.10 \\ 0.12 & 0.76 & 0.05 \\ 0.08 & 0.15 & 0.85 \end{bmatrix}^{40} \times \begin{bmatrix} 300\,000 \\ 120\,000 \\ 180\,000 \end{bmatrix}$

$= \begin{bmatrix} 194\,982.89 \\ 150\,513.16 \\ 254\,503.95 \end{bmatrix}$.

In practical terms, these results are the same as each other. After rounding the numbers, it means that the matrix of the long term number of customers is

$\begin{bmatrix} 194\,983 \\ 150\,513 \\ 254\,504 \end{bmatrix}$.

Solutions: C7

Question 364

a. The set of equations
$$2x + y + z = 12$$
$$x - y + z = 1$$
$$2y - z = 6$$
expressed in matrix form, is

$$\begin{bmatrix} 2 & 1 & 1 \\ 1 & -1 & 1 \\ 0 & 2 & -1 \end{bmatrix} \begin{bmatrix} x \\ y \\ z \end{bmatrix} = \begin{bmatrix} 12 \\ 1 \\ 6 \end{bmatrix}$$

b. The determinant of this matrix (from my calculator's DET command) is 1. Since this is not zero, the matrix is not singular and so there will be a unique solution to the set of equations.

c. If $A = \begin{bmatrix} 2 & 1 & 1 \\ 1 & -1 & 1 \\ 0 & 2 & -1 \end{bmatrix}$, then from my graphical calculator the inverse of A is

$$A^{-1} = \begin{bmatrix} -1 & 3 & 2 \\ 1 & -2 & -1 \\ 2 & -4 & -3 \end{bmatrix}$$

d. Defining matrix B as
$$B = \begin{bmatrix} 12 \\ 1 \\ 6 \end{bmatrix}$$
Then
$$A^{-1}B$$
$$= \begin{bmatrix} -1 & 3 & 2 \\ 1 & -2 & -1 \\ 2 & -4 & -3 \end{bmatrix} \begin{bmatrix} 12 \\ 1 \\ 6 \end{bmatrix}$$
$$= \begin{bmatrix} 3 \\ 4 \\ 2 \end{bmatrix}$$

This leads to the conclusion that the estimated ideal number of bookshops, sports shoe shops and music stores, respectively, will be 3, 4 and 2.

Question 365

a. We don't need the information in the 'energy content' column for this part, and we only need the information contained in the bread and margarine rows. Just copy those six remaining cells.

Answer: $\begin{bmatrix} 1.2 & 20.1 & 4.2 \\ 6.7 & 0.4 & 0.6 \end{bmatrix}$

b. i. First, right down the order of each matrix involved:
$$A \times B = \text{answer}$$
$$(1 \times 4) \times (4 \times 1) = (1 \times 1)$$
$$\therefore \quad \begin{bmatrix} 2 & 2 & 1 & 1 \end{bmatrix} \begin{bmatrix} 531 \\ 41 \\ 534 \\ 212 \end{bmatrix} = [1890]$$

ii. The order of matrix product BA is found as follows:
$(4 \times 1) \times (1 \times 4) = (4 \times 4)$. Answer: (4×4).

iii. The matrix product AB calculates the total amount of energy (kj) one sandwich containing bread, peanut butter and honey and margarine. I note that such a sandwich doesn't really appeal to me.

c.
$$\begin{bmatrix} 1.2 & 6.7 & 10.7 & 0 \\ 20.1 & 0.4 & 3.5 & 12.5 \\ 4.2 & 0.6 & 4.6 & 0.1 \\ 531 & 41 & 534 & 212 \end{bmatrix} \begin{bmatrix} b \\ m \\ p \\ h \end{bmatrix} = \begin{bmatrix} 53 \\ 101.5 \\ 28.5 \\ 3568 \end{bmatrix}$$

$$\therefore \quad \begin{bmatrix} b \\ m \\ p \\ h \end{bmatrix} = \begin{bmatrix} 1.2 & 6.7 & 10.7 & 0 \\ 20.1 & 0.4 & 3.5 & 12.5 \\ 4.2 & 0.6 & 4.6 & 0.1 \\ 531 & 41 & 534 & 212 \end{bmatrix}^{-1} \begin{bmatrix} 53 \\ 101.5 \\ 28.5 \\ 3568 \end{bmatrix}$$

$$= \begin{bmatrix} 4 \\ 4 \\ 2 \\ 1 \end{bmatrix}$$

I used my graphics calculator.
Therefore, $b = 4$, $m = 4$, $p = 2$, $b = 1$.

Question 366

a. Add the elements of the matrix S_0.
$400 + 200 + 100 + = 700$ insects or eggs.

b. Inspect the matrix

$$\begin{bmatrix} 0.4 & 0 & 0 & 0 \\ 0.5 & 0.4 & 0 & 0 \\ 0 & 0.5 & 0.8 & 0 \\ 0.1 & 0.1 & 0.2 & 1 \end{bmatrix} \begin{matrix} E \\ J \\ A \\ D \end{matrix} \text{ next week}$$

The element corresponding to E (eggs) which is followed by E next week is 0.5. That will be the required multiplying factor. Thus, the proportion is 0.5.

c. i. $S_1 = TS_0$

$$= \begin{bmatrix} 0.4 & 0 & 0 & 0 \\ 0.5 & 0.4 & 0 & 0 \\ 0 & 0.5 & 0.8 & 0 \\ 0.1 & 0.1 & 0.2 & 1 \end{bmatrix} \begin{bmatrix} 400 \\ 200 \\ 100 \\ 0 \end{bmatrix}$$

$$= \begin{bmatrix} 160 \\ 280 \\ 180 \\ 80 \end{bmatrix}$$

ii. From the answer to part **i.**, the number of juveniles (the second row) is 280.

iii. $S_4 = T^4 S_0$

$$= \begin{bmatrix} 0.4 & 0 & 0 & 0 \\ 0.5 & 0.4 & 0 & 0 \\ 0 & 0.5 & 0.8 & 0 \\ 0.1 & 0.1 & 0.2 & 1 \end{bmatrix}^4 \begin{bmatrix} 400 \\ 200 \\ 100 \\ 0 \end{bmatrix}$$

$$= \begin{bmatrix} 10.24 \\ 56.32 \\ 312.96 \\ 320.48 \end{bmatrix}$$

This time, the number of juveniles would seem to be 56.32. But we would definitely expect an integer answer, so the result is 56 live juveniles.

iv. I calculated the value of $S_n = T^n S_0$ for various values of n, until the first element (the number of eggs) first became less than one. Look for a very efficient method of doing this on your graphical calculator so that you don't have to re-key the equation over and over. The values I got for $n = 5, 6$ and 7 were respectively 4.9, 1.63 and 0.655. That is, somewhere between weeks six and seven, the number of eggs became less than one. I guess an answer of 7 is acceptable if an integer value of weeks is expected.

v. I started by calculating $S_{20} = T^{20} S_0$

which gave the result $\begin{bmatrix} 10 \\ 0 \\ 11.2 \\ 688.75 \end{bmatrix}$.

Then I tried $S_{70} = T^{70} S_0$ which gave the

result $\begin{bmatrix} 0 \\ 0 \\ 0 \\ 700 \end{bmatrix}$. That would appear to be the

required steady state matrix. However, I would not like to have to work out exactly in which week that matrix first appeared.

d. $S_1 = TS_0 + BS_0$

$$= \begin{bmatrix} 0.4 & 0 & 0 & 0 \\ 0.5 & 0.4 & 0 & 0 \\ 0 & 0.5 & 0.8 & 0 \\ 0.1 & 0.1 & 0.2 & 1 \end{bmatrix} \begin{bmatrix} 400 \\ 200 \\ 100 \\ 0 \end{bmatrix}$$

$$+ \begin{bmatrix} 0 & 0 & 0.3 & 0 \\ 0 & 0 & 0 & 0 \\ 0 & 0 & 0 & 0 \\ 0 & 0 & 0 & 0 \end{bmatrix} \begin{bmatrix} 400 \\ 200 \\ 100 \\ 0 \end{bmatrix}$$

Using a previous result,

Solutions: C7

$$= \begin{bmatrix} 160 \\ 280 \\ 180 \\ 80 \end{bmatrix} + \begin{bmatrix} 0 & 0 & 0.3 & 0 \\ 0 & 0 & 0 & 0 \\ 0 & 0 & 0 & 0 \\ 0 & 0 & 0 & 0 \end{bmatrix} \begin{bmatrix} 400 \\ 200 \\ 100 \\ 0 \end{bmatrix}$$

$$= \begin{bmatrix} 190 \\ 280 \\ 180 \\ 80 \end{bmatrix}$$

ii. $S_2 = TS_1 + BS_1$

$$= \begin{bmatrix} 190 \\ 280 \\ 180 \\ 80 \end{bmatrix} + \begin{bmatrix} 0 & 0 & 0.3 & 0 \\ 0 & 0 & 0 & 0 \\ 0 & 0 & 0 & 0 \\ 0 & 0 & 0 & 0 \end{bmatrix} \begin{bmatrix} 190 \\ 280 \\ 180 \\ 80 \end{bmatrix}$$

$$= \begin{bmatrix} 130 \\ 207 \\ 284 \\ 163 \end{bmatrix}$$

The number of live eggs in this insect population after two weeks is 130.

1. Core. Data analysis. A1, A2, A3, C1.

1.1 Univariate data. Types of data, numerical versus categorical, discrete versus continuous data.

1.2 Presenting univariate data. The various table and graph forms including frequency tables, stemplots, barcharts and histograms. Symmetry, including positive and negative skew of graphs. Remember, a long tail to the left means a negative skew. Outliers should be identified where appropriate. Use of the IQR to identify outliers. Study $Q_1 - 1.5IQR$ and $Q_3 + 1.5IQR$ and how they are used.

© Cambridge University Press

1.3 Summary statistics for univariate data. Measures of central tendency and dispersion: mean (μ and $\bar{x}$), standard deviation (s and σ), median and quartiles, IQR (interquartile range) and range. The use of the 68% and 95% rules for ± 1 and ± 2 standard deviation intervals.

1.4 Boxplots (including those with outliers). Outliers are those items of data outside the 1.5 IQR limits.

© Cambridge University Press

1.5 Bivariate data. The display of the relationships between two variables. Trends between two variables and the notion of independent (x) and dependent (y) variables. Bivariate data displays: scatterplots, back-to-back stemplots, parallel boxplots, and their use to display positive or negative trends. The shape (linear or otherwise) and strength of correlation of a bivariate graph and the estimation of this correlation. Correlation is weak if r is below 0.49, and is strong if $r > 0.75$.

© Cambridge University Press

1.6 Correlation and Pearson's product-moment correlation. The use of a graphical calculator with bivariate statistics to calculate r. Calculating the coefficient of determination, r^2, and the statement that $r^2 \times 100\%$ of the variation observed in one variable can be explained by the variance in the scores in the other. Study how the least squares formula is found – not just from the calculator, but from the original data.

1.7 Regression. Lines of best fit of bivariate data. The methods are 'by eye', three-median and least squares. Interpretation of slope and intercepts of the regression line. Making predictions with extrapolation and interpolation.

© Cambridge University Press

1.8 Residual values, their plots and their interpretation.

1.9 Transformation of non-linear data to linearity by square, square root, log or reciprocal transformation.

1.10 Time series data, plots, trends, seasonal, cyclic and random patterns. How to make seasonal adjustments: seasonal indices, deseasonalisation of data using yearly averages.

1.11 Smoothing: three-median and moving average smoothing with centring where required.

© Cambridge University Press

2. Module 1. Number patterns and applications. B1, C2.

2.1 Recognition of various progressions including the arithmetic, geometric and Fibonacci.

© Cambridge University Press

3. Module 2. Geometry and trigonometry. B2, C3.

3.1 The basic trigonometric ratios and problems involving their use. Similarity. Pythagoras' theorem, basic geometry as applied to regular polygons. The ideas of corresponding, alternate angles and the angle properties of regular polygons.

3.2 Pythagoras' theorem in two and three dimensions.

3.3 Scale diagrams used to represent practical situations.

© Cambridge University Press

4.2 Constructing and interpreting a variety of graphs, including graphs that convert units, show tax scales, bus fares and postal charges.

4.3 The algebraic and graphical solution to linear simultaneous equations in two unknowns. Applications to 'break-even' financial analysis.

4.4 The interpretation of the y-intercepts, slope, the end-points and the average rate of change over graph segments.

© Cambridge University Press

2.2 Sequences and series. Arithmetic progressions. The general term is $t_n = a + (n-1)d$ and the sum of n terms is either $S_n = \frac{n}{2}(2a + (n-1)d)$ or $S_n = \frac{n}{2}(a + l)$, where a is the first term, d the common difference and l the nth term. Geometric progressions. The general term is $t_n = ar^{n-1}$ and the sum of n terms is $S_n = \frac{a(1-r^n)}{1-r}$ and the sum to infinity is $S_\infty = \frac{a}{1-r}$, where $|r| < 1$. a is the first term, r the common ratio.

© Cambridge University Press

2.3 Applications of progressions, including growth models and compound interest.

2.4 Difference equations. Obtaining the terms of a sequence from a difference equation. Interpreting a statement like $t_{n+1} = 2t_n + 3$. The graphical representation of such a sequence. Practical uses of progressions – growth models for example. The use of first-order difference equations to describe change. The notion of initial conditions, usually given as t_0 or t_1. The Fibonacci progression as a difference equation.

© Cambridge University Press

3.7 Interpreting triangulation data.

3.8 Distances and bearings. Calculation of angles and distances in a given 'real world' triangle.

4. Module 3. Graphs and relations. B3, C4.

4.1 Plotting and sketching straight lines graphs. Finding the equation of a straight line in the form $y = a + bx$ from the coordinates of two points or from one point and a gradient.

© Cambridge University Press

3.4 Surface area and volume and the effect on area and volume when you alter dimensions such as r by a factor.

3.5 The cosine rule $a^2 = b^2 + c^2 - 2bc \sin A$ and the sine rule $\frac{a}{\sin A} = \frac{b}{\sin B} = \frac{c}{\sin C}$. The area of a triangle $A = \frac{1}{2}bc \sin A$. Heron's Formula:

$A = \sqrt{s(s-a)(s-b)(s-c)}$ where $s = \frac{a+b+c}{2}$.

3.6 Two-dimensional problems using compass bearings, contour maps and the average slope between two points on such maps.

© Cambridge University Press

4.5 Construction of non-linear and segmented graphs from tables of data. Predicting values for the future from such graphs, estimation of maximum and minimum values.

4.6 Interpreting points of intersection of two graphs and obtaining them on a calculator.

4.7 Interpreting equations of form $y = kx^n$, $n = -2, -1$, 1, 2, 3 and obtaining a linear graph by plotting y against a modified x-variable.

4.8 The constant of proportionality and testing models from a given set of data. The use of interpolation and extrapolation.

4.9 Linear programming. Converting a worded problem into a series of constraints of the form $3x + 2y \le 200$ and finding an algebraic expression for the objective function. The idea of defining variables, listing the constraints and of testing the objective function.

4.10 Graphing linear inequalities. Finding the feasible region and the coordinates of the corners of such regions. Finding the optimum corner and thus solving simple linear programming problems with two variables.

© Cambridge University Press

5.2 Calculating the balances in bank accounts, using information about deposits and withdrawals and the rules for bank interest. Daily, weekly, monthly, quarterly and yearly interest calculations. Such banking problems won't exceed five transactions such as deposits, withdrawals and addition of interest.

5.3 Hire purchase. Flat interest rates and effective interest rates. Hire purchase contracts and their computation.

© Cambridge University Press

5.4 Monetary growth and decay. Using graphs to display money growth or depreciation. Interpreting their shape and meaning.

5.5 The compound interest formula
$A = P \times \left(1 + \dfrac{r/n}{100}\right)$ and its use where A is the amount accrued, P is the principal, r is the interest rate per period, and n is the number of periods.

5.6 Given A, P and r, estimation of the number of payment periods by using a graphical calculator, by means of systematic trial and error or possibly by the use of logarithms and the annuities formula.

© Cambridge University Press

5. Module 4. Business-related mathematics. B4, C5.

5.1 Simple interest. The use of the simple interest formula $I = \dfrac{PRT}{100}$, transposed if necessary. Banking, investments, term deposits, bonds and debentures.

5.7 The comparison of flat rates, reducing balance loans.

5.8 Reducing balance loans. Step-by-step calculation of interest. The manual calculation of the amount still owing after 1, 2, 3, 4 or 5 repayments.

5.9 The annuities formula $A = PR^n - \dfrac{Q(R^n - 1)}{R - 1}$ where P = the original amount, $R = 1 +$ (the interest rate per period), and n is the number of periods. Finding A or Q including in cases where the loan is fully repaid. Use of the annuities formula on a graphical calculator.

© Cambridge University Press

5.10 The total repayment time, and total interest paid, when varying the repayment amount, the frequency of repayments, and the interest rate.

6. Module 5. *Networks and decision mathematics. B5, C6.*

6.1 Undirected graphs and networks. Vertices and edges. Loops and regions. Graphical and adjacency matrix representation of networks.

6.2 Planar graphs and applications of Euler's formula $v - e + f = 2$.

6.3 Complete graphs, paths and circuits. Spanning trees, minimal spanning trees. Weighted graphs where a number is associated with each edge.

6.4 Eulerian and Hamiltonian paths and their applications.

6.5 Shortest paths between two points in a network by informal methods.

6.6 Prim's algorithm for finding minimum spanning trees and applications.

6.7 Directed graphs and networks. The graphical and matrix representation of graphs.

6.8 Critical path analysis. Location of the critical path by forward and backward scanning.

6.9 Network flow. Finding minimum and maximum flows in a network.

6.10 Bipartite graphs. The Hungarian algorithm.

6.11 Words: what are graphs, subgraphs, vertices or nodes, edges, loops, the degree of a vertex, isolated vertex, degenerate graphs, connected graphs, circuit, trees, spanning trees, complete graphs, simple graphs, bipartite graphs, digraphs, Hamiltonian and Eulerian paths and circuits, networks, critical paths, planar graphs, adjacency matrices?

7. Module 6. *Matrices.*

Matrix representation and its application including:

7.1 Matrix representation of data from a variety of situations defined in rows and columns.

7.2 Application of matrix arithmetic including sum, difference, scalar multiple, product, inverse, transpose.

7.3 Simultaneous equations including practical situations, and their solution using the inverse matrix method.

7.4 Using transition matrices to analyse practical situations. Informal consideration of steady state.

7.5 Define the following terms: matrix, vector, scalar, row, column, order, inverse, state, transition, steady-state, game and fair game, optimal strategy. Matrices of low order, with up to five rows or columns, will be used. In SAC activities, tasks and problems in coursework, higher order matrices could be considered.